THE WORLD OF A STREAM

other titles in the series

THE WORLD OF AN ESTUARY
Heather Angel

THE WORLD OF A TREE
Arnold Darlington

also by Heather Angel

YOUR BOOK OF FISHES

THE WORLD OF A STREAM

HEATHER ANGEL
MSc FIIP FRPS

with line drawings by
Christine Darter

FABER & FABER
3 Queen Square London

First published in 1976
by Faber and Faber Limited
3 Queen Square London WC1
Filmset and printed in Great Britain by
BAS Printers Limited, Wallop, Hampshire

ISBN 0 571 10450 9

Contents

Illustrations

PLATES

All photographs, with the exception of plate 10, were taken by the author

FIGURES

Acknowledgements

I am especially grateful to Christine Darter for the care she has taken over all the line drawings. I should also like to thank Dorothy Herlihy for typing the complete manuscript so speedily as well as helping to check it.

Throughout the year, my husband, Dr. Martin Angel, has helped in making notes, collecting samples and identifying specimens from The Stream. He has also read the complete text and made many useful suggestions for improvements.

Finally, my thanks to THB for allowing me free access to his land. Without this, the book could not have been written.

Foreword

A stream is not just a groove in the ground along which water flows, but is a fascinating habitat in which a great variety of plants and animals live. This book describes the life of one stretch of a Hampshire stream which has been studied throughout a whole year. In addition to the resident animals, many visitors come to the stream.

Because the stream runs through farmland, its name and location are not given, and all the names in Figure I are fictitious.

The aim of the book is to encourage readers to discover, explore and to make a study of their own local stream. This book will serve as a guide as to when, where and how to look. Because no two streams are identical, what you see and record will not be exactly the same as I have described on these pages. Indeed, each visit made to a stream will reveal not only new plants and animals, but also new points of interest about familiar ones.

Go out and look for yourselves.

1. Streams in General

No stream is exactly like any other, and no two stretches of a stream are the same. The nature of a stream changes as it grows from a tiny trickle arising from a spring or from a high mountain valley. It may either feed another, or it may combine with several to form a river. A turbulent stream tumbling and cascading down a mountain side is quite different from a slow, placid lowland stream, meandering through peaceful meadows. But all streams do share one common factor—their continuous flow of water. The speed or strength of the water current (the rate of flow) varies not only from one stream to another, but also along different parts of the same stream. The water speeds over the shallow gravel areas or riffles, and idles through the deeper pools. A stream may appear quiet and placid in summer but become an inhospitable torrent after winter storms.

There is no clear distinction between a stream and a river. A river is wider and deeper and in olden days would have presented a considerable problem to a traveller who wished to cross it. Streams are much easier to study as most parts are safely accessible from the banks or by wading.

Quite different names are given to a stream in various parts of the country. In Scotland, a stream is referred to as a burn (Plate 1), in the West Country as a creek. Other commonly used names for a stream include brook and beck (especially for streams with a stony bed or a rugged course). Less well-known names for a stream include branch, brooklet, feeder, fresh, freshet, gill, reach, rill, rillet, rivulet, runlet, runnel, sike, streamlet and torrent.

The common names of some plants and animals describe where

1. A Scottish highland stream or burn tumbles down over rocks in Perthshire.

they normally live and therefore provide the clue as to where they can be found. For example, brooklime (*Veronica beccabunga*) grows in brooks and streams. But sometimes these common names can be misleading. Brookweed (*Samolus valerandi*) grows in wet places near the sea, while pond skaters (*Gerris* spp.), some pondweeds (*Potamogeton* spp.) as well as the river snail (*Viviparus viviparus*) and the river limpet (*Ancylastrum fluviatile*) live in streams as well as ponds and rivers. So throughout this book, whenever a plant or animal is first mentioned, its scientific name will be given afterwards in brackets.

(a) Source of streams

Where a stream begins is known as its source. This source may be a mountain tarn, a crack (or fissure) in limestone rock, a spring bubbling up from underground, water seeping down through boggy ground or melting snow. Snow-melt streams, which are very cold, often cascade down steep rock faces, with no obvious plants growing beside them (Plate 2).

Places where water seeps continuously from the ground are ideal for the growth of certain mosses. Such regions, where mosses cover an extensive area of sloping open ground, are known as a bryophyte flush. Bryophytes is the general name given to mosses and liverworts, which are both non-flowering plants.

(b) Types of streams

The steepness of the slope or gradient down which a stream flows helps to determine its character. Mountain streams fall steeply and the fast-flowing water scours away the soil and eats back into the bed rock. If the bed rock is soft the rush of water erodes out a deep channel. Where water flows over harder rock, no erosion takes place and a waterfall may form. These fast clear streams contain a lot of dissolved oxygen and are the home of the brown trout (*Salmo trutta fario*) and of invertebrate animals, including insect larvae, which are adapted to clinging on to stones.

In very cold springs and streams through Europe where the water

2.
A snow-melt stream in the Spanish Sierra Nevada range, in June

temperature remains, even in summer, below 10°C, live freshwater animals known as *relicts*. They are the survivors of the fauna which occurred throughout Europe during the Ice Age. These rare animals include a flatworm (*Crenobia alpina*), an amphipod (*Pontoporeia affinis*), an opossum shrimp (*Mysis relicta*), a caddis larva (*Apatidea muliebris*) and a water mite (*Spherchon squamosus*).

Further down as the gradient becomes gentler, the stream broadens out and the flow becomes slower. Sand and gravel scoured from higher up are deposited (Plate 3), and even in the pools fine mud or silt settles out on to the stream bed. The water warms up, especially in summer, and so it contains less dissolved oxygen. A sudden storm may result in a rapid rise in the water level. Such a 'flash flood' may burst the banks, flooding surrounding fields. The mud and silt stirred up from the bed of the stream make the water cloudy or turbid.

The speed of the current affects what kinds of plants and animals live in a particular stretch. The banks exert a drag on the flowing water, so that the surface water is drawn in towards the centre where the flow is fastest. This can be seen by dropping a stick in at the edge of a stream. Streams, unlike ponds in which there is little or no flow of water, rarely have floating plants and animals, since they would soon be washed away downstream. However, duckweeds (*Lemna* spp.) and pond skaters can be found on the water surface in quiet eddies and backwaters of streams. In these regions there will also be a wide variety of rooted weeds, including pondweeds (*Potamogeton* spp.) and Canadian pondweed (*Elodea canadensis*). Many of these weeds are completely submerged and so can be identified only by collecting them with a net or a grapnel.

Some water weeds which produce floating leaves are easy to recognise on sight. These include the various water lilies and the common pondweed (*Potamogeton natans*) with its simple oval leaves. The East Anglian fenland dykes are often completely choked with water lilies, including the yellow-flowered brandy bottle (*Nuphar lutea*) and the white water lily (*Nymphaea alba*) (Plate 4).

In fast-flowing streams there is never such a variety of water weeds. But what weeds are present, may produce long trailing stems which snake backwards and forwards as the water flows over them. For example, the willow moss (*Fontinalis antipyretica*), which grows attached to stones, tree roots and pieces of wood in streams, rivers and lakes, is the largest British moss. This aquatic moss can grow up to a metre in length. Its second, specific name, which means 'against fire', arose from its use as an insulating material for packing between

3. A gravel-bedded stream in the New Forest, Hampshire, in April

4. A man-made dyke in Wicken Fen in August, with white water lilies (*Nymphaea alba*) and reeds (*Phragmites communis*)

the spaces of the wooden walls of Swedish houses. Like asbestos, the moss is non-inflammable (will not burn) and it was therefore an insurance against accidental fires.

Some kinds of water crowfoot (*Ranunculus* spp.) grow in fast-flowing streams. Their underwater leaves are finely divided so as to offer least resistance to the water currents. A few floating leaves are produced to support the white flowers above the water surface. Other species which grow in ponds or in slow-flowing streams have more and larger floating leaves.

(c) How streams are fed

We have already seen various ways in which streams begin life. But once begun they must be maintained. The most obvious source of stream water is directly from rainfall (and hail and snow). Much more water runs off the surface of the surrounding land, or seeps in through the banks.

Not all the water in a stream will flow into the main river or into the sea. Some will become soaked up by parched soil along the margins of the stream bank. Water is also lost by evaporation from the surface of the stream or by being drawn up by the roots of green plants and transpired. Transpiration is the evaporation of water vapour from leaves through tiny pores called stomata. Aquatic plants with floating leaves have stomata only on the upper leaf surface. The stomata are prevented from being waterlogged by the waterproof waxy coating on the leaf surface. Each pore is surrounded by a pair of guard cells. These cells regulate the amount of water loss by expanding or contracting to close or open the pore. There are no stomata on the submerged leaves of aquatic plants.

Stream life is not nearly so varied as terrestrial life, but it is none the less interesting. Many of the ways in which freshwater organisms have become adapted to live in a constantly moving underwater environment will be described later in this book.

2. The Stream

When I began searching for a suitable stream to study for at least a year, I decided that it must be within easy reach from my home, so that not too much time would be spent travelling to and fro on each visit. I therefore looked out my local Ordnance Survey 1:50 000 map (this is the new series which is replacing the one-inch maps) and drew a pencil circle with a 30-centimetre radius from my house, which represented about 30 minutes' driving.

I rejected several streams close by, because they were either inaccessible or too overgrown. In the end, the one I chose was 10 miles away and could be reached in 20 minutes. I saw it first late one August afternoon, when the teasels (*Dipsacus fullonum*) and ragwort (*Senecio jacobaea*) were flowering along the banks (Plate 5). As soon as I had selected The Stream, I set to work on finding out as much information as I could about its origin.

(a) Researching the background

Firstly, I decided to visit the local museum to see if The Stream was mentioned in any books in the library. Had it always flowed along the same course? Where was its source? Over what type of ground did it flow? Were there records of any floods? Had it been used as a source of water by local people in the past?

After spending a whole morning delving amongst old books describing the history, the geology and the geography of Hampshire, I not only had the answer to most of these queries, but I also

5. The Stream near Mint Place in August. Teasels (*Dipsacus fullonum*) flowering in the foreground

discovered many other fascinating facts about The Stream and its surroundings.

If you want to find out more about a stream in your area, and there is no museum to visit, try the nearest public library. You could also get in touch with a local historical or natural history society. A Citizens' Advice Bureau will be able to give you useful addresses.

(i) *The Source*

I found many references to the source of The Stream. It is a spring which rises at an altitude of 130 metres (400 feet) at the base of a chalk hill. This spring has never been known to fail. 'On 10 September 1781, after a severe hot summer, following a dry spring and winter, the spring produced nine gallons of water in a minute. . . . At this time, many of the wells failed, and all the ponds in the Vales were dry.'

The spring was marked on the 1:25 000 Ordnance Survey map, which is more detailed than the 1:50 000 map. So I decided to see it for myself. I walked along the footpath running beside the overgrown stream at the edge of a cornfield reduced to stubble. After 100 metres, the green strip of trees and shrubs petered out and I could no longer hear flowing water. Nor could I see the spring. I looked at my map again and checked my position. It *seemed* to be the right place, and yet I could see nothing but dense undergrowth spreading over the steeply sloping ground to the right of the path.

I decided to walk carefully down the slope, prodding the way ahead with a long stick to make sure there were no holes beneath the stinging nettles (*Urtica dioica*). As I descended, the nettles gave way to hazels (*Corylus avellana*) and elders (*Sambucus nigra*) entwined with ivy (*Hedera helix*) and wild clematis or old man's beard (*Clematis vitalba*)—a typical plant of chalk country. The wild clematis stems hung down rather like lianas in a tropical rain forest, so that I had to bend almost double to make any headway at all.

The soggy ground was covered with mosses and pieces of rotting wood on which various fungi were growing. I spotted brown Jew's ear fungus (*Auricularia auricula*) growing on elder. The small white branched spikes of the candle snuff fungus (*Xylaria hypoxylon*) contrasted well against the dark rotten wood on which it was growing. Small groups of the gelatinous purple knot fungus (*Coryne sarcoides*) were also growing on rotten wood. Tufts of the velvet shank (*Flammulina velutipes*) were much more conspicuous with their yellowish brown sticky caps and dark brown stems.

I had expected to see clear water bubbling up from the ground. Instead, I saw a bare-sided depression in which water trickled up through a mass of leaves and twigs. It was Autumn when I visited the spring, and although leaf fall had begun, very little light was penetrating through the overhead canopy. The ground below was so gloomy that I had some difficulty in focusing the camera to take the view shown in Plate 6. The poor lighting meant that in order to get any photograph at all, I had to screw the camera firmly on to a tripod and expose the film for $1\frac{1}{2}$ minutes!

I measured the distance across the spring head depression and noted it to be 90 centimetres. I then walked downstream some 50 metres, where the banks were lined with mosses and ferns, including the hart's tongue fern (*Phyllitis scolopendrium*). At this point, The Stream was 80 centimetres wide. I did intend to visit the spring later during a different season, but I became so absorbed in the life associated with the downstream stretch (which is described in detail in Chapters 3, 4, 5 and 6), that I have not managed to return during this year.

6.
The source of The Stream is a spring which seeps up through a depression filled with leaves and twigs.

One account which I found describes how one day in November, 1951, after 48 hours of a continual heavy downpour—following a wet summer and autumn—water burst forth from the hillside above the spring and poured over the spring across a road below. As The Stream flows eastwards a shorter stream, which sometimes dries out, merges with it. On this November day, below the meeting of the two streams, the water flooded a streamside cottage to a depth of eighteen inches (45 centimetres). During the year I studied The Stream, I did not see anything quite so spectacular, but the pictures of The Stream in spate (Plates 28 and 43, pp. 68 and 95) were also taken in November.

(ii) *The Geology*

While the source and the course of streams and rivers, as well as their gradients, can be seen marked on the Ordnance Survey maps, these maps do not tell you what type of rocks and soils rivers flow through. To obtain this information, you will have to refer to a geological map. Small-scale maps on which the major towns, lakes and rivers are marked are reproduced in the British Regional Geology series of handbooks. These maps are not nearly so detailed as the one-inch to one mile series of coloured Geological Survey maps which are also produced by the Institute of Geological Sciences (see Chapter 8(d)). On these maps each type of deposit is shown as a different colour. The Stream is clearly marked as rising from the Lower Chalk, then as it flows north-eastwards, passing through a band of Upper Greensand, it merges with a smaller rivulet. Still flowing north-eastwards, it passes through a band of Gault into the Folkestone Beds, which form the upper layer of the Lower Greensand. At this point the bed of The Stream is made up chiefly of river sands and gravels (alluvium).

Both the Lower and the Upper Greensand layers were laid down during the Cretaceous period 70–140 million years ago. The Lower Greensand underlies the Upper Greensand and is therefore older and was laid down earlier. The green coloration of many of the sandstones produced during this period gives rise to the name Greensand. When these sandstones become exposed to the air, their green coloration disappears and they turn a rusty brown.

Today, no obvious demarcation line can be seen between these strata where they are covered with vegetation, although west of the stream-head lies an extensive area of chalk which is characterised by typical chalk vegetation, including wild orchids. There is a pleasant walk alongside The Stream, as it flows north-east through a wooded valley, before passing through the stretch of farmland where I worked.

(iii) *The History*

It was a great surprise to discover that the stream selected purely because it was close to my home and its banks were accessible for sampling the water life, turned out to be of such historical interest.

Firstly, the hamlet close beside the stretch where I worked, is mentioned in the Domesday Book; which originally was kept not far away at Winchester.

Then, as the spring head had never been known to fail, the water course close to its source was diverted in 1894 to be the water supply for the village which gives its name to the parish. The overflow at this point spilled out through a lion's head in a wall into a trough below and continues to do so today (Plate 7). As the village population grew, so this water supply became inadequate.

7. The Lion's Head fountain, through which gushes the overflow of The Stream

The stream water was also used several centuries earlier by monks. After the spring-fed stream joins forces with the more northern branch, it runs north-east. A mile past the junction, it passes the site of a ruined priory, which was founded in 1232 by the Bishop of Winchester. The monks used The Stream for their water supply by making a dam from an earth bank. This bank can still be seen today.

(b) Fieldwork

(i) *Preliminaries*

Before undertaking a long-term study of a stream, you should make quite sure that you will not be trespassing on private land. Look on a map to see if a public footpath is marked alongside the stream. If there is no such footpath, find out to whom the land belongs and get his permission to enter his land and to collect any specimens. As mentioned in the Foreword, the stretch on which I chose to work flows through farmland. I therefore telephoned the farmer and arranged to meet him. When I told him about this book he was most interested, as it turned out he was a keen naturalist himself.

With his permission, I was now free to visit The Stream as often as I wished. I aimed to visit it about once a week, but this proved to be impractical, since my work takes me away from home for a week or more at a time. However, since many of the flowering plants flowered for a period of several weeks, I was able to see and to photograph all the common species. But since observing and recording animal

8.
An open stretch of The Stream in early April. Lady's smock (*Cardamine pratensis*) flowering near Pebble Point

behaviour can never be predictable, the more visits are made to an area, the greater will be the chance of seeing something really interesting. This can often occur when you are least expecting it. Patient observation is the only way of making worthwhile discoveries.

Wherever possible observe and make notes in preference to collecting.

Before beginning any fieldwork it is sensible to ask yourself several questions about your proposed plan of study. In this way, your time in the field will be spent most profitably. Firstly, what is it that you want to discover about your stream? If you are more interested in birds than in plants, then you will probably want to compile lists of residents and visitors—month by month throughout the year. Which birds depend on the stream for food or for a nesting site? How extensive are their territories?

On the other hand, if you are more familiar with the plant life, then you will want to study this aspect first. How are the different kinds of plants distributed along the stream banks? In what order do they come into flower? How are the aquatic plants adapted for living in flowing water? Which plants are eaten by what animals?

A useful way of keeping and comparing nature records in general is by making your own nature calendar. Use a diary—it need not be an up-to-date one—and record your observations on the actual date. I have been using the same page-a-day diary for the last six years. I can now see at a glance and compare the dates when a particular fungus first appeared, a plant flowered or set seed as well as when an insect first emerged or a bird began to sing in any of these years. It will also become apparent if a season is an early or a late one.

But whatever aspect you select at first you will soon discover that you cannot ignore the other inhabitants of the stream habitat. For instance, what animal feeds on the neatly cut ends of rushes? What causes the strange red swellings on willow leaves? A good field naturalist looks at a habitat as a whole, and becomes aware of the ways in which the plants and animals constantly interact.

Expensive equipment is by no means essential for making useful field observations. There is no substitute for a keen pair of eyes. However, there are a few basic pieces of equipment which can be extremely useful. Binoculars will enable you to identify birds more quickly by their plumage and shape of beak and so learn their characteristic song or way of flying (see Plate 46, p. 100). 7 × 50 binoculars are powerful enough for most purposes and are not too heavy. A hand lens will reveal details of flowers and seeds which cannot be seen with an unaided eye (Plate 17, p. 53).

You should carry a field notebook so that you can make notes and sketches about your observations *at the time*. It is all too easy to forget or confuse essential details—even later on the same day. You will soon devise your own shorthand; for example, biologists use symbols

9. An overgrown stretch of The Stream in August. View from Brooklime Bend along Teasel Channel

to denote male ♂ and female ♀. I often record my observations on a pocket tape recorder. This is much quicker—especially with numb fingers on cold winter days—than writing notes. A camera is another way of making a permanent record of your observations, especially if you want to check something months later which you considered to be unimportant at the time. Basic photographic techniques are given in Section (iii) below. The relevant specialised field equipment is described at the beginning of Chapters 4, 5 and 6 and a summary of all the equipment can be found in Chapter 7.

(ii) *Mapping*

One of the first jobs will be to make a sketch map of the stream and to label the different parts, so that when writing your field notes, you can refer to these names. Although The Stream was clearly marked in blue on the 1:25 000 Ordnance Survey map, this did not provide enough detail for my purposes. So I had to order a 6-inches to one-mile map on which the bends of The Stream were marked more clearly.

I laid the map flat on a table and photographed it from above. After the transparency was processed, I projected it on to a white wall and made an enlarged sketch map on a piece of cardboard of the stretch on which I intended to work. This technique was used to draw the map in Fig. 1 (see pp. 40–1).

I took the sketch map with me on my next visit to The Stream, and marked in the position of each large tree and the places where the cattle came down to drink. I also named each stretch of The Stream. These names which appear in Fig. 1 are referred to throughout the next four chapters.

(iii) *Photography*

A useful way of supplementing field notes and confirming facts is to take photographs. However, photographs will be of value only if they are sharp and in focus. A fuzzy out-of-focus photograph will not tell you anything. One of the easiest subjects to photograph is a general view of the habitat. Take a series of pictures of the same view in different seasons, as I did with Plates 13, 14, 15, 16 (pp. 45, 47, 49, 51).

If you cannot use a darkroom, then a Polaroid camera will provide you with a print only seconds after exposing the film. However, the cheapest models of this type of camera cannot be used for close-up photography. Close-up pictures must be taken to show the detailed shape and structure of a plant or an animal. Examples of close-up photographs taken in the field can be seen in Plates 18, 19, 22, 24, 26, 49, 51, 52, 53 and 54 (pp. 58, 63, 65, 66, 106, 109, 110 and 113). All these were taken using a single lens reflex (SLR) camera. With this type of camera the subject is brought into focus by adjusting the focusing ring until the image appears sharp on the focusing screen. If a non-reflex camera is used for close-ups the distance between the subject and the camera will have to be measured. This takes longer than focusing by eye and may frighten an insect before you have a chance to photograph it!

The cheapest way of taking close-ups is to use a close-up lens on the front of the camera lens. Extreme close-ups (Plates 37, 38, and 40, pp. 84, 85 and 87) can be taken only by inserting extension tubes or bellows between the camera lens and the camera body. Both extension tubes and bellows are much more expensive than a close-up lens and can be used only with cameras which have removable lenses. For sharp, clear action pictures, use either a 'fast' film and a fast shutter speed (e.g. 1/250 second) or an electronic flash.

Since birds and mammals tend to be wary of a close approach, they cannot easily be photographed using a basic camera with a standard lens. A long focus or a telephoto lens will produce a larger image on your film than a standard lens used from the same position. These longer lenses are heavy, which makes it more difficult to hand-hold the camera without shaking it and thereby blurring the picture. You will get really sharp pictures only if the camera is mounted on a tripod and the shutter operated by using a cable release (Plate 10). Some kinds of telescopes can be used as especially long lenses for photographing very shy birds and mammals, but then a robust tripod really is essential.

Many more detailed photographic techniques are given in my two books listed in Chapter 8.

10.
The author, beneath the oak tree, using a tripod to support the camera with a long focus lens

(iv) *Sampling the water*

Water is essential for all life on Earth, but for freshwater animals and plants it is especially important. Many quickly die if they are stranded or become dried up. Obvious exceptions are insects like water beetles and water boatmen which can fly off and find a new stretch of water. More unusual are the fairy shrimps (*Chirocephalus diaphanus*) which occur in a few temporary pools in Britain. The eggs of these primitive crustaceans must dry out, otherwise they fail to develop. In desert regions, similar animals survive as dormant eggs for years until rain falls and they can hatch.

The quality of the water is important in determining the types of animals and plants that live in it. An obvious example is that very few animals can survive in water heavily polluted with sewage or farmyard water. The breakdown of the sewage by bacteria uses up all the oxygen in the water so that only specialised animals can survive, like the rat-tailed maggot of the drone fly (*Eristalis* sp.) and sludge worms (*Tubifex* spp.).

Several simple tests made on the water itself will provide you with information which may help to explain why some kinds of life are present or absent from a particular stretch of water.

PH The acidity or alkalinity of water (the pH) can be determined by using litmus paper. The dye coating on the paper changes colour when it comes into contact with acids or alkalis, turning red with acids and blue with alkalis. Litmus is known as an indicator substance. Test some litmus paper with lemon juice, vinegar or washing soda and see what colour it turns. Match these colours with the scales on the inside of the pack. There you will find a scale of numbers (pH numbers) alongside the range of colours between red at one end of the scale and blue at the other. These numbers are a quick shorthand way of indicating how acid or alkaline is a solution or a substance. Low pH values indicate acidity, whereas high pH values indicate alkalinity. Pure water is neither acid nor alkaline; it is neutral with a pH of 7.

More accurate pH values can be found by using strips of Universal Indicator Paper. See if your teacher has any at school. Either collect some stream water in a *clean* jar and test it at home or do the test beside the stream. Dip the paper into the water and out again immediately. If it is left submerged the indicator dye will wash out. In a stream which flows from chalky ground the water will tend to be slightly alkaline, with a pH of 8, whereas if it comes from an acid peat bog it will be acid, with a pH as low as 6 or even 5. I found the pH of my stream to be 6·5, using Universal Indicator Paper.

HARDNESS When soap is added to 'hard' water which contains calcium or magnesium salts or a mixture of both, it does not lather easily and it produces a scum. It also causes creamy-coloured deposits ('fur') to build up on the inside of kettles, whereas 'soft' water lathers very easily and leaves no scum.

Since freshwater molluscs (snails, freshwater mussel and limpets) and crustaceans (especially crayfish) require calcium to build up their shells, they prefer to live in hard waters. Therefore if many snails are present in a stream, the water is likely to be hard, alkaline, and to contain an ample supply of dissolved calcium. But if only a few small snails occur, the water is likely to be lacking in calcium and be soft. The degree of hardness can be found by taking a water sample and measuring how much of a soap solution is needed to make a froth. Hard water will require more soap to make it froth than soft water.

LIGHT Streams which flow through woodlands or beneath overhanging alders (*Alnus glutinosa*) will receive less light than those which flow through open meadows with no trees growing along their banks. Sunlight is essential for photosynthesis—the chemical process whereby green plants build up their food from carbon dioxide and water. This process involves steps which are extremely complicated, but in simple terms it is the capturing of the energy given off by the sun as chemical energy, using the green plant pigment known as chlorophyll. The main products of photosynthesis are sugars and oxygen.

Since all plants and animals require oxygen to breathe, the oxygen which is given off by green plants can then be used for respiration by both plants and animals. Aquatic plants, such as Canadian pondweed, are specifically sold as oxygenating plants for garden ponds. Bubbles of oxygen produced by aquatic plants can be seen in still pond water on sunny days.

Comparative measurements of the amount of light reaching the water surface of a stream can be made using a photographic light meter. Hold the meter as close to the water level as possible, with the light-sensitive 'window' looking upwards instead of down into the water. Place the white diffusing cone over the window so that the light falling on the water is measured. Do not point the meter directly towards the sun. Take readings along an open stretch and compare them with readings along an enclosed stretch. Compare the types and the relative amounts of water plants growing in each part.

The amount of light reaching plants underwater is also affected by the clarity of the water. During floods, when water washes off the surrounding land and down the banks, fine mud particles (sediments) carried in suspension will turn the water cloudy or turbid, so that much less light reaches the submerged plants. As the flood subsides, the water will gradually clear.

Streams flowing close to quarries or open-cast mines may have constantly turbid waters. Such streams will have gradually lost many of their aquatic plants, and as a result far fewer animals will live in the stream. The dust clouds which settle on terrestrial plants will also affect them by clogging the pores or stomata on their leaves. Such fine dust, even if completely inert, can be considered to be a form of pollution which is particularly acute in areas where China clay is mined.

Comparative measurements of the amount of silting in a stream or river can be made by collecting a water sample in a small graduated measuring cylinder, allowing the particles to settle and measuring the depth of the sedimentation layer at the bottom.

POLLUTION still occurs all too frequently in our streams and rivers. Effluents from factories may contain poisonous chemicals which kill the aquatic life directly. Insecticides sprayed on to crops may be carried by the wind or washed by the rain into the stream. Herbicides used to control roadside verges or even the water plants themselves may kill many animals as well. Or else rotting vegetation, like sewage, uses up oxygen in the water so the animals die from suffocation. Fish are particularly susceptible to such pollution during warm weather. In regions where the oxygen content has been reduced by pollution, liquid oxygen can be pumped directly into the water. If you come across any evidence of pollution, such as oil slicks, detergent foams, or masses of dead and dying fish, or the mud at the bottom is black and evil-smelling, report your finds to the Pollution Officer or the River Purification Service of your Regional Water Authority. Their address and telephone number can be found under River Authorities in the yellow pages of the telephone directory. You may even be able to trace the source of the pollution by searching up-stream.

TEMPERATURE Even the apparently innocuous warm water effluents from factories can be a form of pollution. The warmer the

11. The Stream gently flowing beneath the last bridge, beside the oak, in August. Compare with Plate 43

water the less oxygen will dissolve in it, and yet the more active are the animals, so they need more oxygen. However, in cold temperatures, animals may become inactive. The larvae of the great diving water beetle (*Dytiscus marginalis*) are inactive and do not feed at temperatures of less than 10°C. Once the temperature warms up above 10°C to about 15°C, they become active—feeding voraciously and growing rapidly.

Make a graph of the changes in water temperature of your stream month by month. Compare the surface and bottom water temperatures. Do temperatures change after rain storms?

CURRENT SPEED The speed at which water flows down a stream varies along different stretches (Plate 11). Objects lying on the stream bed can affect the rate of flow. On the downstream side of stones and boulders where the current is reduced, many small animals take shelter. The rate of flow at the surface can be measured by timing a floating object over a measured distance.

3. Season by Season

(a) A Walk along The Stream

So that you can become familiar with the different stretches and landmarks of The Stream, I will describe the walk along the stretch illustrated in the sketch map of Fig. 1.

You reach the bank of The Stream from the road by climbing over a low wooden fence into a field on the southern side. A few metres downstream is a place where the cattle come down to drink and the steep bank is eroded away into a shallow bay. On the opposite bank is Mint Place, named after the patch of water mint (*Mentha aquatica*) that flourishes there (Plate 5, p. 23). Also on the northern bank are the first trees; an ash (*Fraxinus excelsior*) and two hawthorns (*Crataegus monogyna*). On the corner beside the first bridge where the cattle can cross, there is another hawthorn and a field maple (*Acer campestre*).

Downstream of the bridge is Thistle Patch, where the steeply sloping bank makes it difficult to get down to the water's edge. But on the bend opposite the next ash tree is another cattle-drinking place—on the north bank. At Pebble Point—where the water is shallow and the current faster—an area of pebbles can be seen when the water level is low.

On downstream is a series of bends, which I have called Hawthorn Bends, because the southern bank is lined with a series of hawthorn trees (Plate 12), in amongst which are a couple of ash trees.

The Stream then sweeps round a sharp bend, cutting away the banks, which are further eroded by the cattle coming down to drink. The picture in Plate 50 (p. 108) of the cattle drinking was taken from

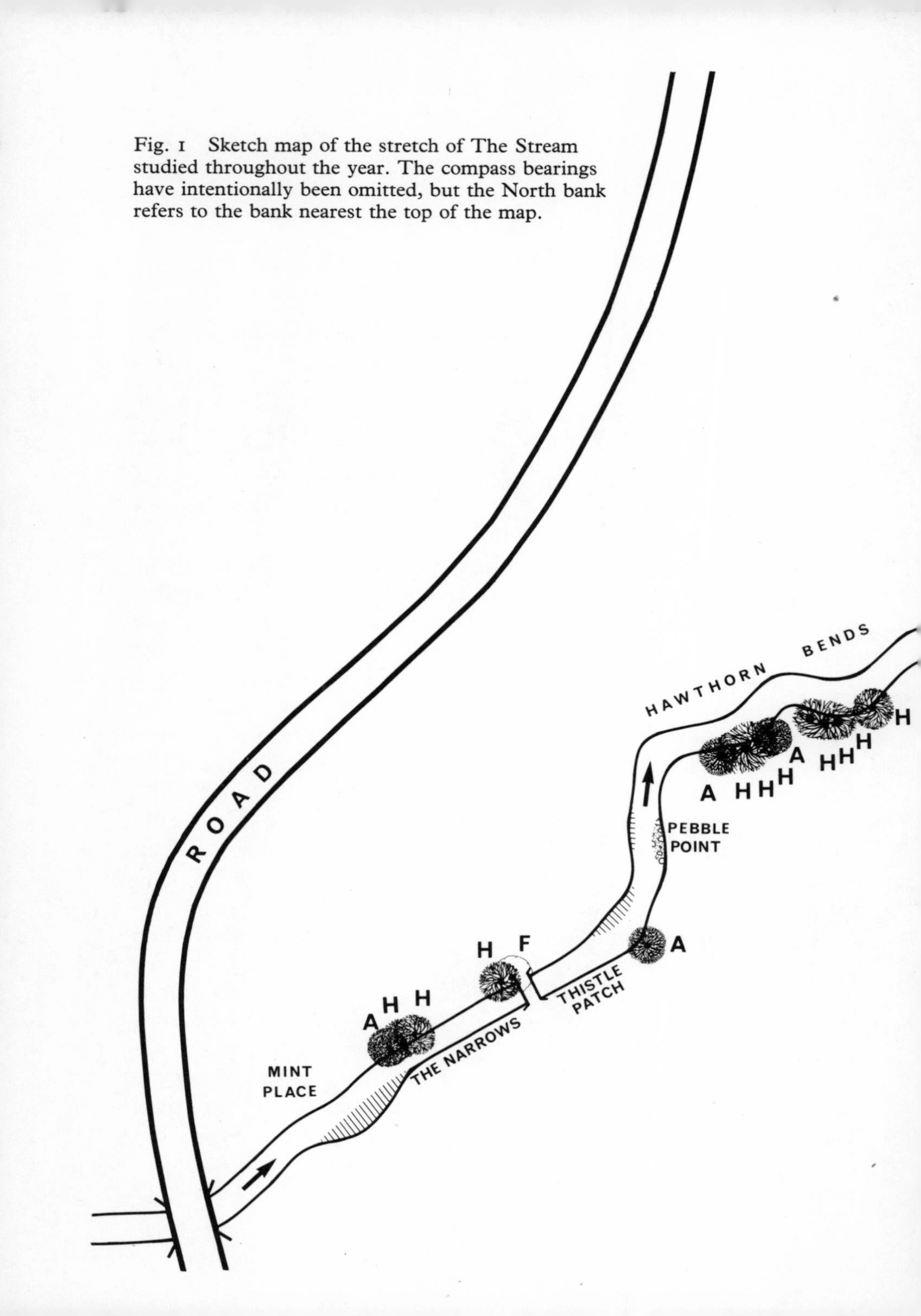

Fig. 1 Sketch map of the stretch of The Stream studied throughout the year. The compass bearings have intentionally been omitted, but the North bank refers to the bank nearest the top of the map.

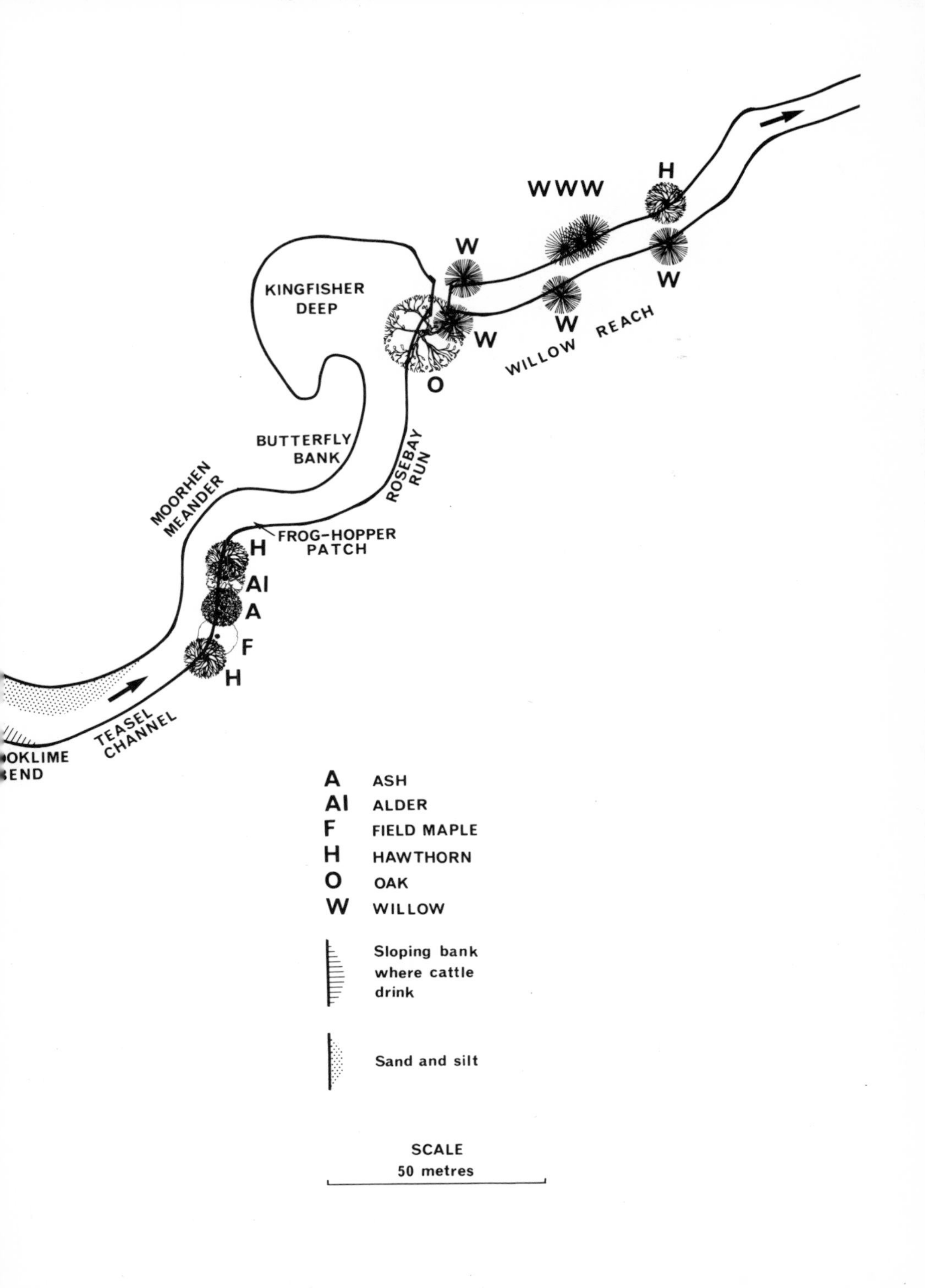

H
WWW
W
W
KINGFISHER
DEEP
W
W
O
WILLOW REACH
BUTTERFLY
BANK
ROSEBAY
RUN
MOORHEN
MEANDER
FROG-HOPPER
PATCH
H
Al
A
F
H
TEASEL
CHANNEL
OKLIME
END
A ASH
Al ALDER
F FIELD MAPLE
H HAWTHORN
O OAK
W WILLOW
Sloping bank
where cattle
drink
Sand and silt
SCALE
50 metres

12. Cattle grazing in a field beside Hawthorn Bends in August. The largest plants lining the bank are codlins-and-cream (*Epilobium hirsutum*).

the north bank near Brooklime Bend. On the inside of the bend is a temporary sand bank, that is deposited as the flood water falls, flattening what remains of the waterside vegetation.

The Stream then flows through Teasel Channel, where the steep banks are lined on either side with teasels. A line of trees which includes two hawthorns, another field maple, an ash and an alder, flanks the south bank leading to Moorhen Meander and Frog-hopper Patch on the south bank (Plate 31, p. 74). Here in May, the vegetation is covered with black and red frog-hoppers (*Cercopis vulnerata*, Plate 49, p. 106). At Moorhen Meander a small area of silt is sometimes to be seen. Compare the pictures of the Meander taken in April and September (Plates 20 and 25, pp. 61 and 66) and see how dramatically it changes during this five month spell.

Below Frog-hopper Patch is Rosebay Run, with Butterfly Bank on the north bank. In summer this is the richest area for butterflies. The Stream then flows into Kingfisher Deep. A kingfisher (*Alcedo atthis*) has a favourite perch on the great oak (*Quercus robur*) (Plate 10, p. 33) that overhangs the water by the bridge. The water flows on down a concrete slope beneath the bridge (Plates 11 and 43, pp. 37 and 95) into Willow Reach. Here the banks are well shaded by the large crack willows (*Salix fragilis*) and so there are fewer flowering plants.

As the crow flies, the stretch is barely 500 metres long and can easily be walked in ten minutes, but even in such a short stretch there is always something new worth stopping to watch and maybe to photograph.

(b) **Spring** March–April–May

Spring is the first season of the year in the Northern Hemisphere. According to the movement of the Earth about the sun, astronomers define Spring as starting on the vernal equinox (21st March) and ending on the summer solstice (21st June). At the time of the vernal equinox the day and the night are exactly the same length and the Earth is positioned so that the sun is directly over the equator at mid-day. As the summer progresses, the North Pole tilts more towards the sun, so that in the Northern Hemisphere, the days become longer and the nights shorter. At the summer solstice this tilt is at its greatest. The sun at mid-day then lies over the Tropic of Cancer, and it is midsummer's day—the time of the longest day and the shortest night—while in the Southern hemisphere, this is midwinter's day—the time of the shortest day and the longest night. In this book, Spring is taken to last from the beginning of March to the end of May.

The first signs of Spring are always welcome—especially if there has been a long, cold winter. Catkins beginning to open, the first celandines (*Ranunculus ficaria*), and primroses (*Primula vulgaris*) flowering and our resident birds starting to sing again are sure signs that Spring has begun. It is a season of great activity. The spring flowers have to bloom before they are overshaded by the leaves bursting out on the trees. Fish spawn, and the birds start to nest so that the young have the rich feeding of Summer to enable them to grow and store enough energy to see them through the leaner months of late Autumn and Winter.

Plate 13 was taken towards the end of Spring in early May. The crack willows are beginning to leaf out and many of the streamside flowers are beginning to flower. The water level has fallen from the winter flood levels, so that it babbles over the exposed pebbles and stones in the stream-bed.

13. The Stream in Spring. Looking along Willow Reach

(c) **Summer** June–July–August

Astronomically Summer begins at the summer solstice (21st June), the time of the longest day, and ends at the autumn equinox (21st September) when again the equator lies directly below the sun so that day and night are equal in length. This is the warmest season of the year, and I take it to last from the beginning of June to the end of August. It is the time when most plants flower and insects are most active. Plant-feeding insects breed during Summer to take advantage of the vigorous growth of their food. Many migrant birds nest in our countryside then, so as to exploit this abundance of insect life to feed their young. Plant growth continues unchecked throughout the Summer despite the efforts of all the insects, so that parts of The Stream bed become completely choked with vegetation. Plate 14 shows Willow Reach in early August. No water can be seen because of the dense growth of the vegetation along the banks. Both the streamside plants and the overhanging willows cast a heavy shade on the water at this point.

14. The Stream in Summer

(d) **Autumn** September–October–November

Measured by the sun, Autumn lasts from the Autumn equinox (21st September) until the winter solstice (21st December). The winter solstice is the time when the North Pole points away from the sun, so that the sun lies directly overhead on the Tropic of Capricorn. In the Northern Hemisphere it is the time of the shortest day and longest night, whereas in the Southern Hemisphere it is the longest day and shortest night. But animals and plants behave more as if Autumn starts in September, and the countryside is firmly in the grip of early Winter by the end of November.

Autumn is the time when our migrant birds leave for their warmer winter quarters. The flowering plants are beginning to look bedraggled, but there is an abundance of seeds and berries that fatten up and are stored by small mammals. On sunny days, bees still forage for the last drop of nectar from the remaining flowers, and a few butterflies and flies lazily sun themselves. Stimulated by the autumn rains, mushrooms and toadstools appear in abundance before the first frosts. The leaves of the deciduous trees begin to change colour in September. Usually by the end of October many of the leaves have blown from the deciduous trees. These falling leaves are an important source of food for many of The Stream's inhabitants. Several types of caddis fly larvae use the leaves or their stalks to build the protective cases for their bodies.

Plate 15, which was taken at 9 a.m. in mid-October, shows the willow leaves already thinning out. The streamside vegetation is beginning to die back and to be laid flat by the first surges of flood water resulting from the September storms. The rush of water is growing now into an angry roar. On a few days in Autumn the dried stalks of the plants were outlined with hoar frost in the early mornings, before the sun emerged.

15. The Stream in Autumn

(e) **Winter** December–January–February

The coolest season of the year lasts, according to the sun, from the winter solstice (21st December) until the spring equinox (21st March), but here I am calling Winter the period from December until February. The cold and wet weather discourages nearly all plant and animal activity. Many animals hide up in snug corners, waiting for Spring. Underground bulbs and tubers are beginning to grow roots and prepare for their quick spurt of Spring growth. When the soil is icy, plants are unable to draw in water. Trees, like all plants, lose water through the tiny pores or stomata (p. 21) in their leaves. Evergreen trees have waxy coatings to their leaves and lose water very slowly. Deciduous trees would lose much more water than their roots could replace and so be damaged or even killed; hence the reason for shedding all their leaves by the onset of Winter.

Plate 16 again shows Willow Reach, but this time one morning late in February after a light snowfall. The water is now more clearly visible, because the streamside vegetation had been trimmed back the previous October after Plate 15 (p. 49) had been taken. The muddy torrent is rushing away towards the bend below the ivy-covered willow. Throughout the Winter much more water flowed down The Stream than during any other time of year, often making it extremely difficult if not dangerous to try to sample the inhabitants in many places.

16. The Stream in Winter

4. Plants along The Stream

(a) How to study them

In many books for plant identification (floras) the months when a plant is in flower or in fruit are indicated by using the numerical order of the months. Thus, March to June is shown as 3–6, March being the third month of the year, and June the sixth. Try using this shorthand method for writing up your own notes.

Mark the position of any trees growing along the stream bank on your sketch map. Note the time the leaves of each deciduous tree begin to open. Which is the earliest species and the latest species? Do the leaves of all specimens of each type of tree open at the same time? Do the trees flower before or after the leaves open? Are the flowers visited by insects?

Wherever possible, try to identify trees and flowering plants in the field, by taking a field identification book out with you. Remember to take a hand lens (Plate 17). If you are uncertain about your identification and you want to refer to more detailed reference books in a library, collect a small twig or stalk with typical leaves and flowers, but only after making sure it is not the only plant of its type growing there. Do not pick a solitary plant. Always use scissors or secateurs when collecting your specimen, and make careful notes about its size and where it was growing.

Botanists carry their specimens in a metal container called a *vasculum*. Providing the day is not hot, large polythene bags are just as good for keeping flowers and plants for short periods. Tins or match boxes are useful for collecting seeds and small fruit.

17.
Examining
a plant with a hand lens

If you want to have your identification confirmed, the specimen will have to be kept, together with a label. Flowering plants and leaves of trees can be preserved by pressing them between sheets of newspaper under a pile of books. To ensure the specimens are completely dried out and so will not go mouldy, change the paper several times. The dried specimens can then be mounted on sheets of white cartridge paper, using thin gummed strips. A systematically arranged collection of dried plants is known as a herbarium, each specimen being labelled as follows:

Common name
Scientific name
Family
Locality with grid reference
Specimen number
Date
Associated plants
Collector
Identified by

Aquatic plants can also be kept by mounting them on paper, but the best way of showing finely branching patterns is to mount them in water. Fill a shallow rectangular dish (a photographic printing dish is ideal) or a baking tin with water—preferably from the stream itself. Cut up sheets of white cartridge paper to fit inside the dish. Immerse one sheet at a time in the dish and float the plant in the water above it. Separate the branches with a paint brush and gently slide the paper with the plant out of the dish. Cover the specimen with a piece of muslin or nylon tights before completely covering the paper with newspaper. Slide another layer of newspaper beneath the cartridge paper and place all the layers beneath a few books on a flat surface. The newspaper should be changed occasionally. As the plants dry out they adhere to the paper. Providing they are kept flat in a folder, they should remain stuck to the paper.

We have already seen how the pH, the light and the temperature can affect the lives of plants and animals. However, the plants themselves can also affect the rate of water-flow. When aquatic plants grow extensively underwater they reduce the water-flow, causing the stream bed to silt up.

(b) **Spring** March–April–May

One of the first signs of Spring was a cluster of yellow coltsfoot flowers (*Tussilago farfara*) high up on the bank just past Pebble Point early in March. I had already noted the hoof-shaped leaves during the previous Autumn. The leaves, which appear after the flowers have died down, have given rise to several local names, including Horsehoof and Foal's Foot. Coltsfoot has been used both as a cough remedy and as a herbal tobacco.

The alders were the first trees to come into flower in March. The overwintering dull purple male catkins gradually turned yellowish as they elongated and began to shed their dark yellow pollen towards the end of March. The tiny female catkins are also dark purple in winter. In Spring they open into tiny cone-like structures (five are shown in Fig. 2). After pollination has taken place, they turn green. They do not mature until the following year, when they become brown woody cones which open to release their seeds. Two old cones can be seen in Fig. 2. Both male and female catkins are produced on the same tree.

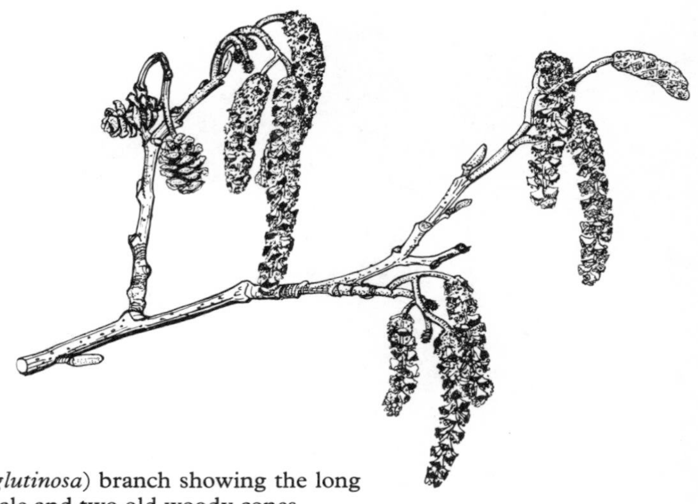

Fig. 2 An alder (*Alnus glutinosa*) branch showing the long male catkins, the tiny female and two old woody cones

Botanists call this *monoecious*, meaning having one home.

At about the same time as the alder catkins begin to develop, the sallow or pussy willow (*Salix caprea*) shoots also burst open their buds. These plants, which were no more than saplings when I first saw The Stream, are not marked on the map, since many were cut back during the Winter. Most of the sallow saplings grow along the northern bank of Hawthorn Bends. Sallow is a *dioecious* type of tree (having two homes) which produces the male and female flowers on different trees. Both types of catkins open before the leaves. On the male trees, the smooth dark-brown bud scales burst open to reveal, at first, the conspicuous silvery hairs. It is amongst these hairs that the mass of stamens covered with yellow pollen develop (Fig. 3).

Unlike many trees, including alder, oak, ash and hazel, sallow pollen is not blown by the wind from the male to the female flowers. Sallow flowers attract a host of insects early in the year when few other plants are flowering. Pussy willow is always associated with Easter, so next Easter have a look at some sallow trees and see how many different types of insects you can see feeding on the flowers. As the insects feed on the male flowers, some of the sticky pollen adheres

Fig. 3 Goat willow or sallow (*Salix caprea*) catkins. The male twig is on the left and the female on the right.

to their bodies and is carried with them when they fly to the greenish female flowers (Fig. 3). So sallow is an example of an insect-pollinated tree. Lime (*Tilia* × *europaea*) is another tree, which in July produces, at the same time as the leaves, its sweet-smelling flowers which are visited by honeybees.

After the sallow has been pollinated, the fruits begin to develop, so that by the end of May they split to release their white fluffy seeds, which are blown around by the slightest suggestion of wind. Seeds falling into the water get carried downstream, where some may become wedged in the bank and, if conditions are suitable, germinate and grow.

The large crack willows did not flower until well into April, when the leaves began to open. These trees, which are typical of stream and river banks, may begin life from a broken branch or twig being carried downstream, becoming lodged in the bank and taking root like a cutting.

Earlier in April the black buds on the ash trees swelled, and turned a dark red colour as they opened before the leaves. On large ash trees, the buds are usually so high up from the ground that it is difficult to see the colour changes which take place as the flowers open and produce their pale yellow pollen. Ash trees are unusual in the way they produce their flowers. Some trees produce only male flowers, some produce only female flowers, others may be chiefly male- (or female-) producing trees with the odd branch producing the other kind of flowers. To add to the confusion, a branch may produce male flowers in one year and female flowers the next!

Oak trees also flower about the same time as the ash—hence the well known saying:

> If the oak be out before the ash
> The summer will be but a splash.
> If the ash be out before the oak
> The summer will be all a soak.

Unlike the ash, the oak flowers open *with* the leaves. The single oak tree along The Stream came into flower slightly after the ash trees.

The hawthorns and the field maple, however, did not flower until the middle of May. The strongly scented creamy hawthorn or may flowers attract insects—especially bees. The pale green field maple flowers produce nectar which attracts small bees and flies. The name maple tree comes from the Old English *mapultreow*. Both in the Midlands and the South, there are examples of place names which are derived from the early names for the maple, including Mapledurwell in Hampshire and Mapledurham in Oxfordshire.

During April I found many horsetail plants (*Equisetum* sp.) with their small cones along the damper parts of The Stream bank. Horsetails are non-flowering plants which have jointed stems with whorls of smaller jointed stems (Fig. 4). Horsetails are often confused

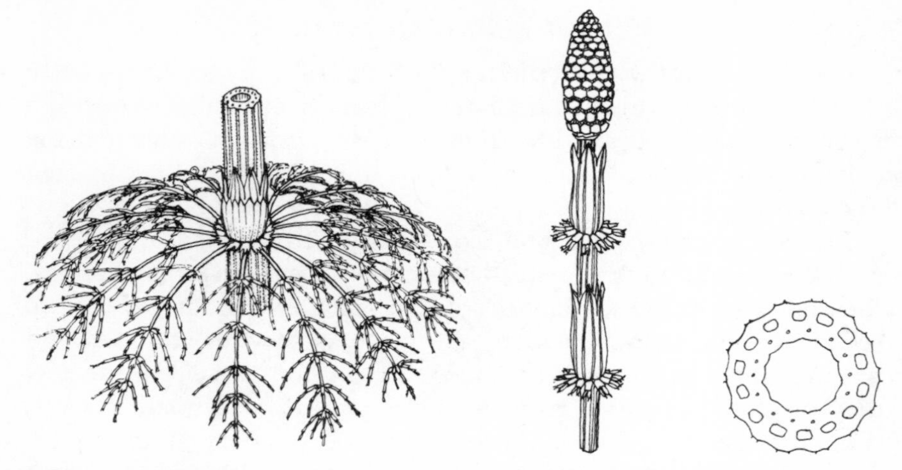

Fig. 4 Horsetail, *Equisetum* sp., showing whorled arrangement of the jointed branches, the spore-producing cone and a section through the stem

with mare's-tail (*Hippuris vulgaris*), an aquatic flowering plant, which superficially resembles horsetails in that it produces its leaves in whorls. Horsetails reproduce by means of spores which are produced by the cone (Fig. 4). They are common as fossils in coal measures and so are much more ancient than flowering plants.

Also in April, some of the flowering plants are beginning to show colour. Early in the month I found dandelions (*Taraxacum officinale*), daisies (*Bellis perennis*), jack-by-the-hedge or hedge garlic (*Alliaria*

18. (*left*) Water forget-me-not (*Myosotis scorpioides*) flowering along Willow Reach
19. (*right*) Brooklime (*Veronica beccabunga*) flowering at Brooklime Bend

petiolata) and lady's smock (*Cardamine pratensis*) flowering. Lady's smock (Plate 8, p. 28) is also known as cuckoo flower, because it begins to flower when the cuckoos are returning. It has pale mauve flowers, and the leaves can be eaten in salads. The seed pods are eaten by caterpillars of the orange tip butterfly (*Anthocharis cardamines*).

Towards the end of May, I found the first aquatic plant in flower—water forget-me-not (*Myosotis scorpioides*) (Plate 18). Soon after, brooklime (*Veronica beccabunga*) (Plate 19) began to bloom. Both have blue flowers and are perennial plants which continue growing year after year. Brooklime was once used as a drug plant; an infusion of the leaves was added to boiling water, and taken to ward off scurvy and impurities of the blood.

Also in May, the conspicuous tall flower spikes of plicate sweet grass (*Glyceria plicata*) (Fig. 5) appear. This grass, which grows in wet places, has a creeping rootstock. On one occasion I noticed a cow eating the leaves and later I observed that anywhere the cattle were able to reach clumps of this sweet grass, the plants had been grazed by them. Tall spikes of winter cress or yellow rocket (*Barbarea vulgaris*) were conspicuous along the length of The Stream at the end of May. These plants, with their mass of small yellow flowers, grew close to the water's edge.

Fig. 5 Plicate sweet grass (*Glyceria plicata*)

(c) **Summer** June–July–August

Throughout Summer, but especially in June and July, The Stream banks take on a range of colours as successions of plants come into flower.

In June, on the drier ground at the top of the banks, bird's foot trefoil (*Lotus corniculatus*) and yellow meadow buttercups (*Ranunculus acris*) were flowering. The sepals (beneath the yellow petals) of this species of buttercup are not bent away from the petals as in the bulbous buttercup (*R. bulbosus*). As I walked through the vegetation, stems of cleavers or goosegrass (*Galium aparine*) caught on my clothes. The tiny hooks on the cleavers' stems and leaves help it to scramble over other plants. When I glanced at the stems I saw that they were covered with tiny whitish flowers. The most conspicuous plants along the tops of the banks at this time of year are the bluish-purple flowers of meadow cranesbill (*Geranium pratense*), the marguerites or ox-eye daisies (*Chrysanthemum leucanthemum*) with the yellow centres to their white flowers, and red campion (*Silene dioica*). Other names for red campion include Soldier's Buttons and Bachelor's Buttons. The latter name arose from a custom dating back to the 16th century and earlier, when girls pinned the flowers beneath their aprons so as to entice the love of the man they desired.

In amongst the hawthorn trees along Hawthorn Bends the elders were flowering in June. The creamy flattened flower heads are strongly scented and are used for making elder-flower wine. In Austria, elder-flower pancakes are made by dipping the flower heads into batter and then frying them.

Beside the water, clumps of wood club-rush (*Scirpus sylvaticus*) were in full flower. The plants die down each Winter, but once they begin to sprout from their rootstocks, they grow apace and by the Summer reach up to 100 centimetres in height. In Plate 20, taken in April, the wood club-rush leaves are only a few centimetres tall, but by June the rush is in full flower and in Plate 21 the tops of the leaves with the flower spikes can be seen.

20. (*left*) View from the bank above Frog-hopper Patch in April, looking towards Moorhen Meander. Wood club-rush (*Scirpus sylvaticus*) growing up on the bend
21. (*right*) Wood club-rush (*Scirpus sylvaticus*) flowering beside Frog-hopper Patch in July

Also in June comfrey (*Symphytum officinale*) (Fig. 6) was flowering beside the water and on the sloping banks; some of the flowers are white, others are pinkish. Comfrey is still used in country districts as a poultice.

Fig. 6 Comfrey (*Symphytum officinale*) in flower

Fig. 7 Hemlock (*Conium maculatum*) showing the mass of tiny flowers arranged in umbels and the blotchy stem

Later on in July, ragwort, yarrow (*Achillea millifolium*), creeping thistle (*Cirsium arvense*), spear thistle (*C. vulgare*) and brambles (*Rubus* sp.) were flowering in profusion on the tops of the banks, while on the sloping edge, several plants of hemlock (*Conium maculatum*) were flowering (Fig. 7). These striking plants can be recognised by the purple blotches on the green stems. All parts of the plant are poisonous. Hemlock is in the same family as cow parsley (*Anthriscus sylvestris*) and parsley (*Petroselinum crispum*), and cases of poisoning are probably due to the plant being mistaken for parsley. Socrates died from drinking a draught of hemlock juice, and Keats in his poem *Ode to a Nightingale* refers to its effects:

My heart aches, and a drowsy numbness pains
My sense, as though of hemlock I had drunk.

22. (*left*) Meadowsweet (*Filipendula ulmaria*) in flower in July
23. (*right*) The spiky fruits of bur-reed (*Sparganium erectum*)

Along the stream banks were the flowering spikes of yellow-flowered square-stemmed St. John's wort (*Hypericum tetrapterum*); cream-flowered meadowsweet (*Filipendula ulmaria*) (Plate 22) and purple loosestrife (*Lythrum salicaria*). As the cattle came down to drink, they ate the meadowsweet leaves. Purple loosestrife tends to grow in open places, because its seeds will germinate only in direct sunlight.

In amongst plicate sweet grass along the water's edge near Froghopper Patch, branched bur-reed (*Sparganium erectum*) flowered. Bur-reeds produce striking spiky globular flower heads (Plate 23).

Rushes grow in clumps like grasses and sedges, and so they are often mistaken for them, but they are grouped together in their own family, the Juncaceae. Rush flowers show similarities to lilies. However, the word 'rush' is misleadingly included in the common names of plants such as bulrush and spike-rush, which are, in fact, sedges (Cyperaceae).

Fig. 8 Three common rushes seen growing along The Stream banks
Left: Jointed rush (*Juncus articulatus*)
Centre: Hard rush (*Juncus inflexus*)
Right: Soft rush (*Juncus effusus*)

In July there were three different species of rushes flowering along The Stream, each of which is illustrated in Fig. 8. Soft rush (*Juncus effusus*) appears to produce a cluster (a panicle) of greenish brown flowers halfway up the straight stem. The part above the flowers is not a true stem but a bracteole, an appendage of the flowers. The long stems of this rush are used for making mats and chair seats, and the pith inside the stems was once used as wicks for candles. Hard rush (*J. inflexus*) produces a loose panicle of greenish brown flowers. The pith does not run continuously down the stem. Jointed rush (*J. articulatus*) can be recognised as one of several species of jointed rushes by running a thumb and finger down the length of the hollow leaves and feeling the cross partitions inside. The chestnut brown flowers are produced almost at the top of the stem.

In August, the most striking flowering plants dotted along The Stream banks are teasels. The order in which the tiny purple flowers open on the flower head is unusual. At first, the flowers in the middle of the head open to form a central band of colour. As these die, more flowers open above and below this region, so that two lines of colour gradually move away from the middle portion (Plate 24). The hooked prickles on the stem and the bracts prevent cattle from feeding on the plants. A variety of this teasel which has hooked bracts on the flower head is still used to bring up the pile on blankets in a Northumberland mill. Rows of the dried heads are mounted on a framework, over which the woven blankets are gently pulled.

Not so tall, but still conspicuous by its mass of yellow flowers, is fleabane (*Pulicaria dysenterica*). The strong smell given off by the leaves deters cattle from grazing on the plants. A smouldering bunch of fleabane is supposed to asphyxiate fleas.

24.
Close-up of a teasel flower (*Dipsacus fullonum*) growing along The Narrows in August

25. (*left*) The same view as in Plate 20, taken in September. Codlins-and-cream (*Epilobium hirsutum*) has grown almost right across The Stream.
26. (*right*) Detail of codlins-and-cream flower (*Epilobium hirsutum*)

In mid-summer large stretches of The Stream became choked with codlins-and-cream or great hairy willow-herb (*Epilobium hirsutum*) (Plate 25). 'Codlin' was an old name for a type of apple. Since apples are often eaten with cream, the rosy petals with their white stigmas (Plate 26) may have suggested the name codlins-and-cream. Other local names for this plant include Cherry-Pie, Love-Apple, Sugar-Codlins and Custard-Cups. The flowers are relatively small compared with the size of the plant. A related plant, rosebay willow-herb or fireweed (*E. angustifolium*) grows in the drier regions along the banks of The Stream. The name fireweed arose from the way the plant spreads rapidly on to burnt areas—especially forest clearings where trimmings are burned. Both types of willow-herb produce seeds with fluffy parachutes which help to carry them in the wind over long distances into new regions.

Another plant which has spread rapidly—especially along waterside margins—is Himalayan balsam or policeman's helmet (*Impatiens glandulifera*) (Plate 27). This plant, which is a native of the

Fig. 9 Detail of water figwort flower (*Scrophularia aquatica*)

27. Flowering spike of Himalayan balsam (*Impatiens glandulifera*)

Himalayas, has escaped from gardens and in some places is the dominant plant alongside streams and rivers. Although I found it in several places along The Stream, it was never as abundant as codlins-and-cream. Himalayan balsam is an annual plant which fires out its ripe seeds when the mature capsule suddenly and violently splits open. Any seeds which land in the water float away for long distances downstream.

Water figwort (*Scrophularia aquatica*) is another common waterside plant. Like all figworts, it has square stems and opposite leaves. Through a hand lens, the small flowers appear quite beautiful both in colour and shape (Fig. 9). Although a variety of insects visit water figwort, it is usually pollinated by wasps (Plate 51, p. 109).

The peppermint scent produced by water mint (*Mentha aquatica*) identifies its presence long before the plant comes into flower in late Summer. The large patch which grows at Mint Place produces masses of tiny pink flowers crowded together in the dense flower heads which are visited by hordes of butterflies (Plate 52, p. 110) and bees. Gipsies use the leaves of water mint to make a type of tea.

Peppermint (*M.* ×*piperita*) is a hybrid plant produced by crossing water mint with green spearmint (*M. spicata*)—a plant cultivated in

Fig. 10 Fool's watercress (*Apium nodiflorum*) is sometimes mistakenly eaten as watercress.

herb gardens. Peppermint is grown commercially and the oil or essence extracted by crushing and heating the crop.

Besides Mint Place and in several other shallow parts of The Stream, fool's watercress (*Apium nodiflorum*) (Fig. 10) grows along the water's edge. This plant is sometimes mistakenly eaten as watercress (*Rorippa nasturtium-aquaticum*), but fortunately it produces no ill effects.

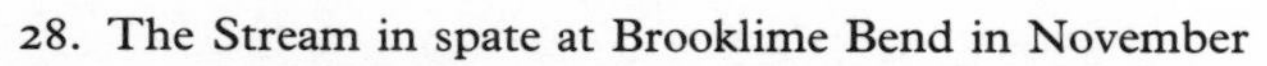

28. The Stream in spate at Brooklime Bend in November

When the water level was low during the Summer, I could see that there were no submerged water plants growing anywhere along my stretch of The Stream. Upstream much of the land is clay, so that any sharp rainstorm produces a rapid spate of flood water (Plates 28 and 43, see pp. 68 and 95). These sudden and isolated fluctuations in the water level probably prevent the establishment of the water plants which are such a common feature of many lowland streams.

(d) **Autumn** September–October–November

The rapid growth made by plants during Spring and Summer begins to slow down as Autumn approaches. By September, most of the annual and perennial plants have set seed and begun to die back. Even so, I found several plants of both brooklime and water forget-me-not still flowering during this month. A large patch of brooklime had been grazed by cattle at Brooklime Bend during the previous month but it had re-grown so quickly that there were no signs of its having been eaten two weeks previously.

Walking through the verge along the top of The Stream, my trouser legs became covered in small hooked fruits of cleavers, reminding me of how I had found the flowers in the Summer (page 60). Several of the local names, such as Huggy-me-close, Sweethearts and Sticky Willie, obviously refer to the way the fruits cling so effectively to clothing. Several other plants produce hooked fruits or seeds which stick readily to animals or humans so that they get widely distributed.

In August the clusters of elder fruits began to ripen and by September they were completely black. Like the flowers, the berries are often collected for wine-making.

In September, the big oak by Kingfisher Deep was laden with acorns. They grew on long stalks, thus distinguishing this English or common oak from the sessile oak (*Quercus petraea*), which has stalkless or sessile acorns. Another difference to look for is at the base of the leaves. The common oak leaf has a lobe on each side of a short leaf stalk, whereas the sessile oak has no lobes and a distinctive leaf stalk.

Once the oak had shed its leaves in October, I found several hard brown marble-sized woody spheres on the twigs. These were oak marble galls and are one of many different types of galls which occur commonly on oak. A gall is an abnormal growth on a plant, usually caused by the presence of either an insect or a mite. Marble galls are caused by a small gall wasp called *Andricus kollari*. Inside each gall a single grub develops which eventually eats its way out to emerge as a type of wasp. See if you can find the exit hole in old marble galls. On common oaks only female wasps emerge which can lay eggs in the oak bud without needing to mate. The species also occurs on turkey oaks, on which both males and females emerge and mate normally. This particular gall was introduced into Devon well over 100 years ago, as a source of tannic acid, which was used at that time for making ink and dyeing cloth.

Another well-known gall found on oak trees is the larger soft, spongy oak apple, which develops in June. This browny-pink gall is also caused by a gall wasp (*Biorhiza pallida*), but instead of only a single wasp emerging from one gall, several dozen wasps will emerge. If galls without any holes are collected early in June and kept in a ventilated jar, the wasps can be seen emerging.

Flattened oak spangle galls are caused by several species of gall wasps (*Neuroterus* sp.). They can be found on the underside of oak leaves in Autumn.

Galls also occur on field maple leaves. Many red pointed projections first appear in June, but they are most conspicuous in early Autumn when the leaves have turned yellow. There may be as many as 500–1000 on a single leaf. These galls are caused by a gall mite called *Eriophyes macrorhynchus*.

The field maple, like sycamore (*Acer pseudoplatanus*), to which it is related, produces winged fruits, each with a horizontal wing. The wing helps to disperse the seed further afield as it falls from the tree on windy September days.

Along Willow Reach many leaves of the crack willows had conspicuous red swellings in the Autumn. These were bean galls, which are caused by a sawfly (*Pontania proxima*). They project equally on both sides of the leaf (Plate 29), and are caused by a

29.
A crack willow (*Salix fragilis*) leaf edge-on, showing bean galls caused by the sawfly *Pontania proxima*

substance injected by the female sawfly when she lays her eggs. The galls are well developed before the sawflies hatch out. Long before the fly is ready to emerge, the larva bites a hole in the bottom of the gall, so that its droppings can escape.

As Autumn drew to a close only a few battered ragwort plants were still flowering in mid-November.

(e) **Winter** December–January–February

Throughout the Winter The Stream banks were bare and lifeless but this bareness revealed several points of interest that were not readily seen during the more verdant seasons.

For example, once the willows had shed their leaves, several plants could be clearly seen growing up their trunks. Ivy (Plate 16, p. 51) was the most conspicuous plant using the willows for support. Ivy grows up from the ground holding on to the trunk, using tiny adventitious roots which grow out from the stem. Ivy flowers in Autumn, but there were still a few of the yellowish-green flowers to be found at the beginning of December. Both wasps and flies visit ivy flowers and pollinate them. Ivy also grows over the walls of the bridge beside the oak tree (Plates 11 and 43, pp. 37 and 95).

Growing in the forks of the willow branches were polypody ferns (*Polypodium vulgare*). Plants which grow on tree trunks, without feeding from them, are known as epiphytes. In tropical rain forests, many plants, including orchids, are epiphytic, so that they receive more light than down at ground level.

Several rotten branches were broken off the willows during Winter gales. Branches which fall into The Stream are gradually disintegrated by the force of the water. Trees are an important element in the landscape, and it was quite a shock to find one day that a gale had toppled the first of the three willows on the north bank of Willow Reach (Plate 30). It had fallen right across The Stream so that the upper branches lay beside the middle willow, very close to where I stood to photograph Plates 13, 14, 15 and 16 (pp. 45, 47, 49 and 51). This particular tree was near to the bridge, but similar fallen trees can become a crossing place for terrestrial small mammals which otherwise cannot cross easily from one bank to the other.

Early in January, I found Jew's ear fungus growing on a piece of elder along Hawthorn Bends. This brownish gelatinous fungus has a velvety texture. It always grows on elders, and can be found during several months of the year—especially in mid-winter.

Both the common and the scientific names arose from the resemblance of the fungus to a human ear and to the legend that Judas Iscariot hanged himself from an elder tree. The original name of 'Judas' ear' has gradually become changed to 'Jew's ear'. This fungus was once used medicinally both as a poultice for inflamed eyes and also as a gargle for throat inflammations; the gargle was prepared by stewing the fungus in milk.

In January, I noticed several clumps of beech (*Fagus sylvatica*) leaves in eddies along the banks. No beeches grew anywhere near the

30. Wind-felled crack willow (*Salix fragilis*) in Willow Reach

stretch I was studying, so the leaves must have been carried down from further up stream, possibly from the Hangers—the local name for the beechwoods that grow on the chalk slopes.

At the end of February, a light snowfall transformed the appearance of The Stream and its banks (Plate 16, p. 51).

5. Life in The Stream

(a) Looking at the water life

A net is the most useful piece of equipment for studying water life. It can either be bought from a biological supplier or you can make your own. The frame should be rigid and either square (Plate 31) or triangular. Pliable circular frames are not suitable because they are not rugged enough. A convenient size is for each side of the frame to measure 25 centimetres. Anything larger will be unmanageable in a current. Use a strong nylon mesh such as curtaining, with a mesh size of 2–3 millimetres. If the mesh is too large, insect larvae, beetles and snails will escape through the holes; if it is too fine, not only will it become easily clogged with silt, but it will also be difficult to handle in a current. However, if you want to catch the small planktonic animals in mid-water, a much finer mesh will be needed. Firmly attach the net

31.
Examining net contents near Froghopper Patch in November

to a stout pole about $1\frac{1}{2}$ metres long; broom handles make suitable poles.

A white or cream-coloured flat-bottomed dish (such as a pie dish or washing-up bowl) is a useful container for turning out the catch to examine it. Forceps or tweezers can be used to pick out small animals, but the smallest ones are best extracted with a pipette fitted with a rubber teat. Fish are best handled using the little nets which aquarium suppliers use for cleaning out their tanks. A small glass trough is ideal for examining specimens in the field, from all angles. This mini-aquarium can be made from small pieces of glass sealed together with the special glass adhesive which is available from pet shops. Alternatively, a temporary trough can be made by clamping two sheets of glass on either side of a piece of rubber or polythene tubing curved into a U-shape, by using two bulldog clips. Sheets of Perspex are less fragile, but soon become scratched. Write notes about the animals as you catch them, only bringing home for identification one or two examples of each in screw-topped glass jars or snap-topped plastic containers. Keep the predators (anything with large jaws and many of the bugs which have piercing sucking mouthparts) separate from the other animals.

Once you start using the net, you will soon begin to find that different animals live in different parts of the stream. Many avoid the main current by sheltering amongst the weeds or roots along the edge, or under stones on the bottom. The more effort you expend in working the net through the overhanging vegetation (hence the need for a robust frame) the greater will be your catch. Along gravel reaches where you can wade safely (Plate 3, p. 18), hold the bottom edge of the net tight against the bottom with the mouth facing the current. Then either turn stones over by hand, or kick the gravel up, just upstream of the net mouth. Animals sheltering in amongst the stones are swept into the net. Many of them, however, are attached firmly and will be found only by careful examination of the rocks.

Weeds can be collected from deep water by using a metal grapnel on the end of a tough rope. Take care that no one is standing too close when you swing the grapnel into the water. Many animals will fall out of the weed if it is drained over a dish, but, as with twigs and leaves from the bottom, closer examination by immersing in a bowl of water will reveal a much greater selection.

In your collecting containers segregate specimens from each microhabitat, writing on each container where the animals came from with a water-proof felt pen. Note the depth the animals came from, estimate the current speed and any other ecological information. It is amazing how valuable a comprehensive field notebook is when you start to try to understand the processes in your stream. Estimates of how abundant the more common, easily recognisable animals are will show up their seasonal variations. But beware of changes in the way you catch the animals; there may seem to be more of them because you learn how to catch them. Similarly, an apparently rare animal might be much more common if you knew where and how to find it.

Do not take a lot of livestock home, unless you can keep it in an aquarium or in large bowls. Unlike pond life, stream life will not stay alive for long, unless the water is aerated with a small aquarium pump. Cover the container with a sheet of glass or Perspex to prevent beetles from flying, bugs from crawling, or fish from leaping out. Keep the aquarium or bowl in a cool place. It is better to keep a few animals to study their behaviour in detail and then return them to the stream and collect one or two new specimens, than to keep a variety all at the same time. The second part of Chapter 7 tells you how to go about setting up a stream habitat.

The majority of freshwater species can be found in some form throughout the year, although certain stages in the life history can be found only for a limited part of the year. So the animal life found in The Stream could be described under any of the four seasons. But since the species breed at specific times, or are more abundant during one season than another, each will be described during this particular season.

(b) **Spring** March–April–May

On land, Spring is the breeding season for many animals. Similarly underwater, many fish are also preparing to breed, although their activities are not so easily seen. The most common fish we found in The Stream were bullheads or miller's thumbs (*Cottus gobio*), three-spined sticklebacks (*Gasterosteus aculeatus*), minnows (*Phoxinus phoxinus*) and stone loaches (*Noemacheilus barbatulus*). Any time from March onwards, female bullheads with their abdomens distended with eggs can be found. Female sticklebacks mature slightly later in the year. We found ripe female bullheads and sticklebacks towards the end of May.

Bullheads are bottom-living fish with a large flattened head and a very wide gape to their mouth. They have no swim bladder, and so sink down to the bottom as soon as they stop swimming. This can be observed if they are kept in an aquarium. Bullheads are crepuscular (active at dusk and dawn) and nocturnal fish, emerging to feed when the light is poor—including dull days. Their mottled brownish coloration blends in with stony bottoms (Plate 32) and so helps to camouflage them. This type of patchy coloration breaks up the body outline, for the same reason buildings are painted with irregular areas of colour during wartime to disguise their shape from enemy aircraft. Bullheads are able gradually to darken or lighten their body colour so as to blend in more effectively with different-coloured stones or gravel. See how long a bullhead takes to change its colour to match its background.

32. Bullhead or miller's thumb (*Cottus gobio*) lies well camouflaged on a gravel bottom

Except during the breeding season, bullheads live a solitary life. They feed on freshwater shrimps (*Gammarus* sp.), mayfly, stonefly, blackfly and caddisfly nymphs (nymph is the name given to an aquatic larva) as well as the eggs and fry of other fish, such as trout.

During the courtship which precedes spawning, the male becomes much darker. The female lays about 100 yellowish sticky eggs in a hole excavated beneath the underside of a stone. The male, not the female, guards the eggs for 3–4 weeks. Once they hatch, the fry scatter from the nest. Bullheads live for 3–5 years and reach a length of 10–18 centimetres.

Three-spined sticklebacks or tiddlers, as they are often called, are widespread in both ponds and streams (except hill streams) all over Britain. These fish have three spines along the top of their back—the front two spines being much larger than the third (Plate 33 and Fig. 11, see pp. 78 and 79). In Plate 33, the spines are lying flat and therefore are not so obvious.

Three-spined sticklebacks can migrate down estuaries into the sea and round the coast of North Scotland they are quite common marine fish. In fresh water, their food consists of worms, molluscs, various nymphs, freshwater shrimps and water lice (*Asellus aquaticus*).

33. Three-spined stickleback (*Gasterosteus aculeatus*) in an aquarium

Young sticklebacks feed on small crustaceans, such as water fleas and copepods.

For most of the year, sticklebacks are yellowish-brown, with pale undersides, but in Spring the males develop a spectacular breeding dress. The belly turns bright red, the back darkens and the ring around the pupil of the eye turns a brilliant blue. This coloration serves two purposes during the breeding season. It is used both for defending his territory against other males and for attracting a mate.

He first builds a nest on the bottom from pieces of weeds stuck together with a secretion from his kidneys. Having done this, he then courts any female fish with a swollen abdomen. She responds by following him and together they perform a zigzag dance (Fig. 11). He shows her the nest entrance and she enters to lay her 100–400 eggs. He then moves in to fertilise the eggs. One male stickleback may succeed in luring several females to lay in his nest. Even then, his work is not over. He guards the nest from attackers and fans a stream

Fig. 11 Courtship display of the three-spined stickleback (*Gasterosteus aculeatus*). The male (upper left) displays his red belly to the female. When she enters the nest to spawn, he follows behind her to fertilise the eggs.

of fresh water over it until the eggs hatch, about a week later. He continues to guard the young fry for the first week or so of their life. If danger threatens a straying young fish, he sucks it up into his mouth and spits it back amongst the rest of the brood.

As the young sticklebacks grow and become more active, they begin to move away from the nest site and find cover amongst vegetation or beneath stones. Sticklebacks become mature after one year and continue to live for a further two years, when they may reach 10 centimetres in length, but are more often only 6 centimetres.

Sticklebacks are eaten by kingfishers, herons (*Ardea cinerea*) and otters (*Lutra lutra*) as well as pike (*Esox lucius*) and perch (*Perca fluviatilis*) (Plate 55, p. 117). They used to be caught in large quantities in German lagoons for their oil and for making into fishmeal.

Minnows, which are found in almost all streams, are very variable in colour. They are usually darker above and paler below, with a distinct stripe running along the sides from the head to the tail (Plate 34). From May to June, minnows collect together in spawning shoals

34. A shoal of young minnows (*Phoxinus phoxinus*) swimming in an aquarium

over gravel bottoms. During this time of year, the males sport red bellies with dark throats and white tubercles on the head. The females lay 200–1000 eggs in clumps between stones. The eggs hatch after 5–10 days and the fish mature after a year. Young minnows feed on water fleas and chironomid larvae (red bloodworms). The larger fish eat stonefly and mayfly nymphs. In Winter they move into deeper water. Since minnows are so widespread, they are an important source of food for other fish and fish-eating birds, and they are also used as bait by anglers.

Like the bullhead, the stone loach is a nocturnal fish which lives on the bottom. It uses its six barbels or feelers to locate its food, consisting of bottom-living organisms, including chironomid larvae, mayfly and stonefly nymphs. All barbels are on the upper jaw, four in front and one at each corner of the mouth (Plate 35). The fish have no scales and are a mottled yellowish-brown colour.

Spawning takes place in April to May, when the eggs are laid in batches in amongst stones and water plants. Young males develop spawning tubercles on the inside of their paired pectoral fins on the underside of the body. These tubercles are present all the year round on older fish, but are most conspicuous in the Spring. Stone loach eggs take about two weeks to hatch and the barbels begin to appear when the fish are about five weeks old. The fish can continue growing until they reach 12 centimetres. Stone loaches are sensitive to

35.
Head of a stone loach (*Noenacheilus barbatulus*) showing the six barbels

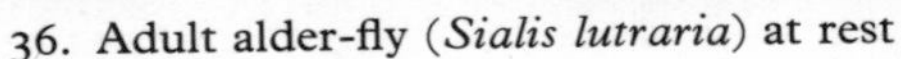
36. Adult alder-fly (*Sialis lutraria*) at rest

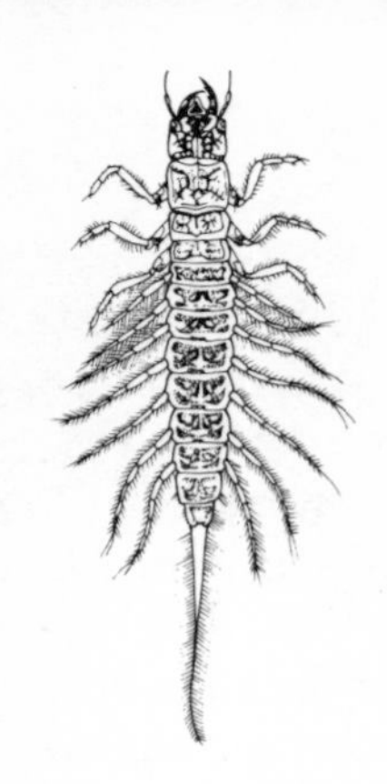

Fig. 12
Alder-fly (*Sialis lutraria*) larva showing the abdominal gills used for breathing, behind the three pairs of legs

pollution and so they will not be found in any polluted streams or rivers. They are eaten by larger fish, pike, trout and eels (*Anguilla anguilla*), by dippers (*Cinclus cinclus*) and by otters.

As the alder leaves opened in May, alder-flies (*Sialis lutraria*) began to emerge (Plate 36). These flies, which are related to the green lacewing flies which come into houses in the Autumn, have brown wings with obvious black veins. They also have a conspicuous pair of antennae and, like all insects, three pairs of legs.

Alder-flies spend most of the time resting on or crawling over waterside vegetation. They are easy to catch even if they are disturbed, since they fly for only a short distance before they settle again. When resting, the wings become folded so as to form a ridged roof across the body. Adult alder-flies do not feed at all, so their life is short-lived. After mating, the female lays up to 2000 brown cigar-shaped eggs in compact batches on plants overhanging the water. When the larvae hatch out they fall or crawl into the water, where they live for a year or more.

The brown larvae have seven pairs of cream-coloured pointed gills which project upwards and backwards along both sides of the abdomen (Fig. 12). There is also a long tapering gill projecting from the end of the abdomen. These gills are used for breathing under water. Alder-fly larvae spend most of their time crawling around on the bottom, but they can swim through the water by undulating their bodies up and down. If the oxygen content of the water is low, the larva remains stationary and undulates its body to produce a current of fresh water over its gills.

Unlike the adults, the larvae are voracious feeders, using their pincer-like jaws to seize midge larvae, mayfly and caddis nymphs. When fully grown, the larva crawls out of the water until it finds a suitable place to dig into some earth. Here, it sheds its larval skin and pupates. This stage in the life history of an insect is known as a pupa. After 2–3 weeks the adult fly emerges.

Alder-fly larvae are eaten by fish—especially trout. Since the larvae feed on other larvae which eat plant material, they are an important link in the freshwater food chain.

(c) **Summer** June–July–August

Sampling life in The Stream was much easier in early Summer when the water level was the lowest it had been all year. It was then possible to wade along most of The Stream, except through Kingfisher Deep.

As already mentioned, there are no water-weeds in The Stream; but at the beginning of June, fine growths of green algae appeared on the muddy margins. Later in the Summer, yellowish brown layers of microscopic diatoms developed on the surface of the mud. Both these plants provide a source of food for the minute larval stages of many freshwater animals.

A very abundant animal found in The Stream throughout the year is the 2–3 centimetre long freshwater shrimp (*Gammarus* sp.) (Plate 37). But in Spring and Summer the females were found being carried around by the larger males 'in tandem'. Also at this time of year, the females were carrying eggs or young in their brood pouches. Freshwater shrimps are a type of crustacean known as an amphipod. Amphipods have many pairs of legs on a body which is usually flattened sideways. The sandhoppers found amongst the strandlines on the seashore are also amphipods.

Freshwater shrimps spend most of their time crawling around

37.
Freshwater shrimp (*Gammarus* sp.) showing how the body is flattened from side to side

amongst stones and leaves on the bottom of streams and ponds. They use their front legs for walking and their back legs for swimming through the water. *Gammarus* is a scavenger which feeds on the remains of plants and animals—known as detritus. Freshwater shrimps moult about ten times before they become mature. In warm summer weather they moult much more frequently (every 5–7 days) than in cold winter weather.

Another scavenging crustacean found in The Stream is the water louse or hog slater (*Asellus*) (Plate 38), but it is not nearly so abundant as *Gammarus*. *Asellus* was found only in the quieter parts where there were no strong currents. Water lice are isopod crustaceans which, like woodlice, have a flattened body. They creep over the bottom and climb up water plants. They can also swim through the water. The larger male settles on the back of the female for about a week prior to mating, when they lie with their undersides together. The female lays about 50 eggs in her brood sac, in which the young are also carried. The sac appears as a large white swelling on the underside near the head end. If part of the antennae (feelers) or legs are damaged, they can be regrown.

38.
Water louse
(*Asellus aquaticus*)
crawling over
a twig underwater

Water lice are easy to keep in a small container with some water weeds. *Gammarus*, on the other hand, will not keep well unless the water is aerated.

When The Stream was first sampled in August, two kinds of freshwater snails, a lesser water boatman (a corixid) and also several leeches were found. These animals were frequently encountered throughout the year.

The larger of the two snails found was the wandering snail (*Limnaea peregra*) (Plate 39). It has a tiny point at the top of its 11-mm-high shell, which has a large final whorl. In late Spring and early Summer, the gelatinous egg masses of this snail were found on the submerged leaves of plants rooted into The Stream bed. If the snails are kept in an aquarium, the development of the embryos inside the eggs can be seen with a hand lens.

The other snail, although much smaller (5mm), was present in much greater numbers. Jenkin's spire shell (*Potamopyrus* (*Hydrobia*) *jenkinsi*) (Fig. 13) is dark brown or black in colour. It is of particular interest for two reasons. Firstly, it was confined to salt and brackish

39. Wandering snail (*Limnaea peregra*) crawling over weeds in an aquarium

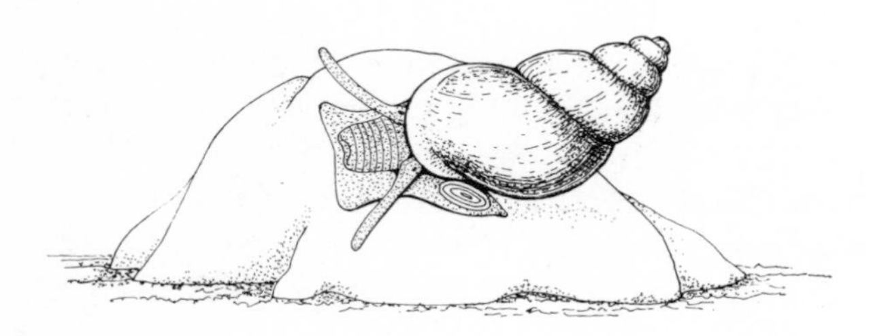

Fig. 13
Jenkin's spire shell (*Potamopyrus jenkinsi*) crawling over a stone. The length of the shell is 3 mm.

waters until the end of the last century, when it began to invade freshwater habitats. Since then it has spread rapidly. Secondly, the eggs develop inside the female without fertilisation taking place (i.e. they develop parthenogenetically) and 35–40 young snails are born alive. Animals which bear their young alive are known as viviparous. Jenkin's spire shell is one of only four species of British prosobranchs (operculate molluscs) which are viviparous. Therefore if only a single snail gets into a pond, ditch or stream, it will soon be able to produce a new colony, because it is able to breed throughout the year. Owing to their small size, these snails are easily carried from one stretch of water to another on birds' feet.

Until recently, no males were known, but one specimen has been found in the Thames in Berkshire. Jenkin's spire shell feeds on detritus, and is related to the estuarine snail *Hydrobia ulvae* which is so abundant on the mud flats and in the creeks at low tide.

Water bugs are a group of insects that includes the large backswimmer (*Notonecta* sp.) and the lesser water boatmen (*Corixa* spp.). They have three pairs of legs and the larger species have piercing mouthparts, used for sucking juices from their prey, which are quite capable of piercing human skin. At the front end of the head is a rostrum (a kind of beak) ending in a fine point. In many species the rostrum is tucked beneath the head and is not easily seen. The corixids feed by sucking up small pieces of detritus and the contents of algal cells, instead of piercing prey.

Both *Corixa punctata* and *C. sahlbergi* were collected from The Stream. They use their middle pair of legs for clinging on to objects under water (Plate 40). If they let go or stop swimming, they will bob up to the surface tail end first, to renew their air supply. Their entire life cycle takes place under water. Nymphs emerge from the eggs, looking like small replicas of the adults without fully developed wings. Each time a moult occurs the wings grow a bit bigger.

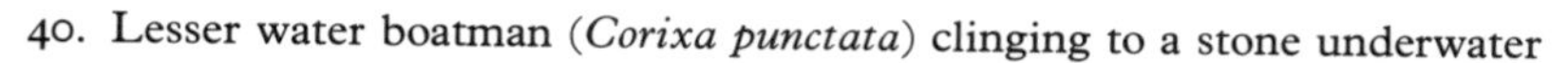
40. Lesser water boatman (*Corixa punctata*) clinging to a stone underwater

On the undersides of stones we frequently found leeches. These are a kind of worm with a sucker at each end of the segmented body. Leeches can readily elongate or contract their bodies, so it is difficult to measure them accurately. Fig. 14 shows a drawing of *Glossiphonia complanata*, which was the most abundant leech in The Stream. At rest, it measures more than 15 mm long. The colour of this leech is very variable, but it is usually dull green or brown with yellow warts. There are also distinct stripes which run down the length of the leech. The three pairs of eyes are usually arranged in two parallel rows.

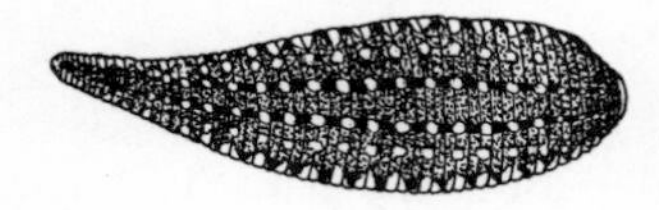

Fig. 14 Leech (*Glossiphonia complanata*) at rest

Leeches have a reputation for being blood-suckers. But in this country, the medicinal leech (*Hirudo medicinalis*) is the only leech which feeds on mammalian blood. For centuries, these leeches were used by doctors for blood letting and so many were collected that they are now comparatively rare.

Glossiphonia complanata feeds by sucking up the body juices of freshwater snails. It lays its eggs in a transparent cocoon on water-weeds or stones. Most of the young leeches reach maturity and breed when they are a year old. They breed for a second time when they are two years old; then they die.

Erpobdella octoculata was another leech found in The Stream. Its upper surface is blackish, with yellow markings. It feeds on aquatic larvae, oligochaetes (worms) and water fleas, by swallowing them whole. *Erpobdella* lays a colourless lemon-shaped 3–4mm-long egg cocoon on objects underwater, which later flattens and turns brown.

The fish leech (*Piscicola geometra*) (Plate 45) also occurred in The Stream and is described on page 97. Leeches move either by looping over objects using their suckers or by swimming through the water.

In August I saw three kinds of insects on the water surface in quiet eddies near the last of the Hawthorn Bends. The most conspicuous were the whirligig beetles (*Gyrinus marinus*), which swim in rapid circles rather like model speed boats. No matter how many beetles there are on a patch of water, they never collide with one another. Like all insects, they have three pairs of legs. It is the short middle and hind pairs with their hairy fringe which are used for swimming. The shiny steely-black wing cases repel water and therefore remain dry.

Whirligigs can dive very efficiently, so the only way to catch them is to scoop them up into a net as quickly as possible. When they dive, they carry an air bubble at their hind end down with them. Their eyes are divided into an upper and a lower part, so when they are on the surface they are able to see efficiently both above and below water. The upper part is used for viewing above water and the lower part for underwater vision.

Whirligigs use their long pair of front legs for seizing food, the insects that fall on to the water surface. The beetles spend the Winter hibernating in mud at the bottom. In Spring, they emerge to lay their eggs in rows on underwater plants. Curious-shaped larvae emerge, which are superficially like alder-fly larvae with long abdominal gills. They use their pointed, hollow jaws for sucking juices from other larvae or even plants.

At the end of July the larvae crawl out of the water to pupate in a cocoon on a waterside plant. The adult beetles emerge in August and are then most abundant. They are able to fly from one stretch of water to another, and also to escape from an uncovered aquarium!

Also on the water surface of the quiet parts of The Stream were pond-skaters (*Gerris* sp.) and the water cricket (*Velia caprai*) (Fig. 15). Both these animals are water bugs which spend nearly all their life on the surface. Female pond-skaters submerge to lay their eggs on

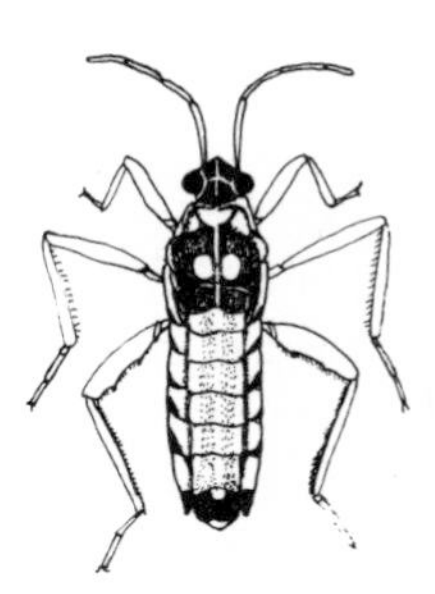

Fig. 15 Water cricket (*Velia caprai*) is a bug which lives on the water surface

41.
Pond-skater (*Gerris* sp.) resting on the water surface. The highlights are reflections from the dimples made by the legs resting on the water.

plants below the surface. Both insects use the surface tension of the water to stay afloat (Plate 41). Pond-skaters are kept dry by air becoming trapped in a layer of fine hairs on the underside of their long thin bodies. They feed on insects which drown on the surface. Female skaters often carry the males, which are smaller, on top of their backs.

The water cricket is broader and shorter than pond-skaters. The body is dark brown with two orange lines along the back of the adult. Most of the adults are wingless, and so will not fly out of an aquarium, but they are very efficient climbers. They, too, feed on corpses. The female lays her eggs either on floating vegetation or on plants close to the banks. It is always worth looking carefully for surface insects and for recently emerged alder-flies, stone-flies, mayflies, damsel-flies or dragonflies on the waterside vegetation, before rushing to start fishing. By approaching carefully, and moving slowly forward, you will see much more.

In the Summer, adult mayflies of the species *Baetis rhodani* were abundant on the waterside plants. Mayflies are particularly graceful insects, with three long tail streamers and delicate veined wings which, when the insect is at rest, are held closed together above the body. The general name given to the group to which mayflies belong is Ephemeroptera, from the Greek *ephemeros*, meaning lasting only a day. These insects are unusual in that a sub-imago (the dun of the fly fishermen) emerges from the last larval stage, flies up from the water surface and immediately moults again to form the adult (or spinner). Most adult mayflies live for only a few hours or a day or two at most. But when a mayfly 'hatch' takes place, the trout greedily feed on them, taking full advantage of this abundance of food. Fly fishermen mimic the aquatic stages with their 'wet' flies and the adult stages with their 'dry' flies.

Baetis rhodani produces two generations in a year and the adults continue to emerge over a long period. The nymphs were found throughout the year. Like the adults they have three tail filaments (caudal cerci), and in addition, they have a series of leaf-like gills projecting along the sides of the abdomen (Fig. 16).

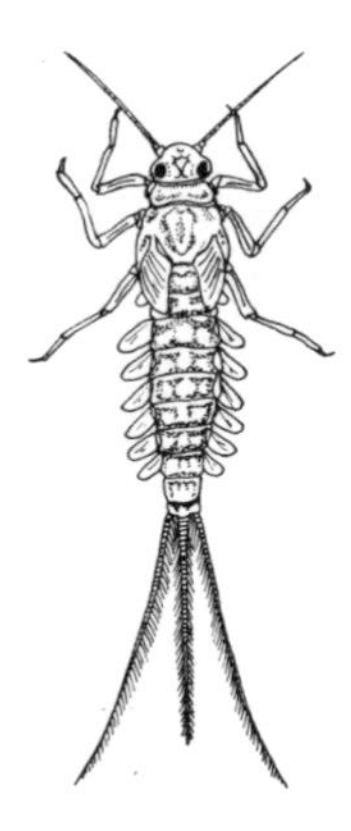

Fig. 16 Mayfly (*Baetis* sp.) nymph showing leaf-like abdominal gills and three long tail filaments

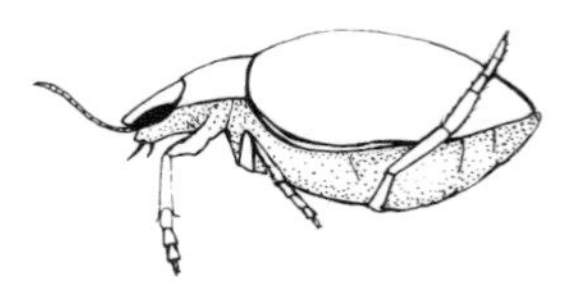

Fig. 17 Globular water beetle (*Hyphydrus ovatus*) is only 5 mm long

(d) **Autumn** September–October–November

In addition to the whirligig beetles, living on the surface, there were two smaller aquatic beetles underwater. *Hyphydrus ovatus* is a 5-mm-long reddish-brown globular beetle (Fig. 17) which has no common name. Like all beetles, it passes through four distinct stages in its life history. The larva which emerges from the egg eventually turns into a pupa, from which the adult emerges. *Hyphydrus* larvae reach 7–8 mm long and so are longer than the adults. The larvae are supposed to be present in the Summer, but we did not manage to find any in The Stream.

Also in September, we found many tiny riffle beetles (*Limnius volckmari*) beneath stones. These black beetles are only 3 mm long. Except for the pupal stage, the whole life cycle takes place in the water. Larvae leave the water in August to pupate 10–15 centimetres from the edge of the water. Some larvae which do not pupate in the Autumn, overwinter for a second time. Therefore both adults and later larval stages were found throughout the year.

Riffle beetles belong to a family now known as the Elminthidae (it used to be known as Elmidae or Helmidae). All stages of the life history have a tough cuticle. The beetles cannot swim, so they crawl about on the bottom, clinging on to stones by long sharp claws on the ends of their legs.

Several species of case-building caddis larvae were found in The Stream. These larvae are difficult to identify unless they are removed from their cases and examined under a low-power microscope. Case-building caddis larvae are examples of animal architects, each species using a consistent design and usually one kind of material to construct its protective tube. Cases are made from pieces of reeds, leaves or twigs as well as from sand grains, small stones or even small empty shells.Both caddis and mayfly larvae provide an important food source for many fish.

When I visited The Stream early one Autumn morning, I disturbed a common frog (*Rana temporaria*) (Plate 42) resting amongst the dew-covered grass. Having a moist skin, frogs prefer damp shady habitats on land. Like most amphibians, frogs must return to water to breed. Usually they select a shallow pond, rather than streams, for spawning.

As well as the leeches, other smaller worms live in The Stream. There are the thin flatworms which creep over the surface of submerged stones, leaves or weeds. Unlike leeches, they have no

42. Common frog (*Rana temporaria*) head-on

suckers and the body is not segmented. In shallow streams where the current is small, flatworms can be collected by dipping a piece of raw meat, attached to a piece of string, into the water. Preferably choose a clear stream so that you can watch the black or brown flatworms converging on to the meat. Dangle it beside the edge of large stones. Transfer some flatworms to a pie dish and watch the way they glide over the bottom and also upside down beneath the water surface. They move using tiny whip-like organelles called cilia, which occur in vast numbers over the whole body surface. All the ten British species of triclad flatworms are carnivorous. Small animals are sucked up whole by a tubular sucking pharynx on the underside. Larger prey (or the meat portion) is covered in slime and pieces are sucked off it. Flatworms lay their eggs in small round cocoons which are attached to stones or water-weeds.

After several days of heavy rains in mid-November, the level of water in The Stream rose two metres to just below the top of the banks (Plates 28 (see page 68) and 43). Since The Stream runs through a deep channel the sudden rise in water level altered the whole atmosphere. Compare the surging water in Plate 43 with the tiny trickle which flowed beneath the bridge in August (Plate 11, p. 37). After these heavy rains fell, large portions of the banks became eroded away by the flood waters.

(e) **Winter** December–January–February

Owing to both hibernation and migration, the variety of aquatic life collected during Winter is less than during the warmer months, but many of the animals described earlier in this chapter overwinter and therefore can still be found in the water. One kind of larva which overwinters is the blackfly (*Simulium* sp.). For most of the year, this insect passes unnoticed, but it is all too apparent when the females emerge and start biting! The species which occurred in The Stream was *Simulium ornatum*, the commonest of 35 blackfly species which occur in Britain.

Both the larvae and the pupae seek out places where the current is strong. They can be collected by lifting up stones and examining

43. The Stream in spate beneath the last bridge, beside the oak, in November. Compare with Plate 11

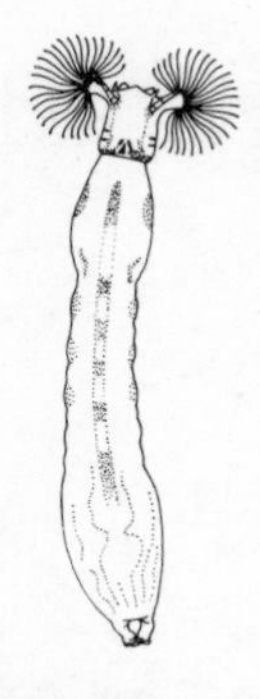

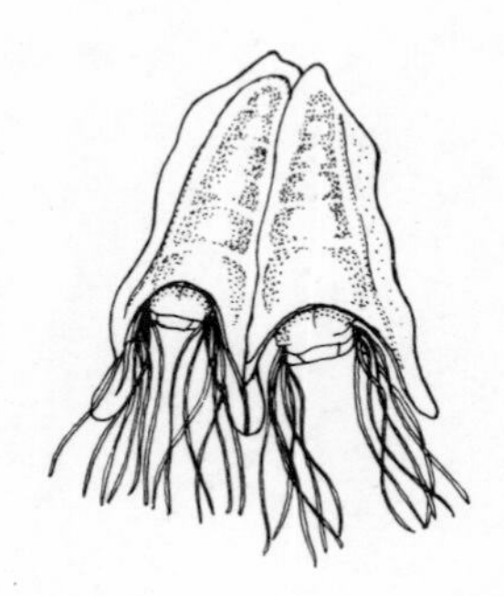

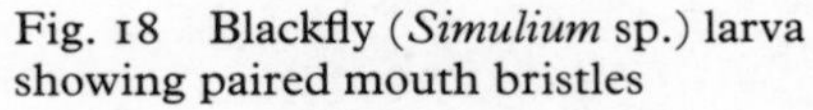

Fig. 18 Blackfly (*Simulium* sp.) larva showing paired mouth bristles

Fig. 19 Two blackfly (*Simulium* sp.) pupae with their filamentous gills

weeds. The dirty cream larvae cling on to stones or weeds by means of a sucking disc at the end of the abdomen (Fig. 18). They also have a smaller sucker near the front end, which is used for moving in a looping fashion. At the head end is a pair of conspicuous mouth bristles which are used for combing small particles (algal cells and detritus) from the water current. Towards the end of Winter, the larvae begin to pupate. They spin conical brown pupal cases either on the stones or on plant stems which are open at one end (Fig. 19). From this end, a pair of branching filaments project forwards and act as gills.

When the adult fly is ready to emerge, it becomes surrounded by a film of air, which buoys it up to the surface from where the fly flies off. Female *S. ornatum* suck the blood of farmyard animals—especially cattle. In some areas, blackflies are a great pest. One such region is Speyside in Scotland, where they are known as birch flies.

When the level of The Stream was low, and the water was clear, we saw a small brown trout swimming upstream. The top of its body was pale brown with paler sides, with red and black spots. Trout are a prized sporting fish, and so young fish are reared in trout hatcheries to stock rivers and lakes.

Winter is the season when the wild trout population breeds. The female digs a depression (a redd) on a gravel bottom by flexing her tail back and forth. As she sheds her orange eggs, the male moves in close beside her, to fertilise them with his white milt. The eggs take over a month to hatch, the exact time depending on the water temperature. The stage which emerges from the egg is known as the alevin. On its underside is the orange yolk sac which provides it with food for the first few weeks. Once the yolk has been absorbed, the fish—now known as fry—feed on small aquatic organisms, like mayfly and beetle larvae, and chironomid worms. The fry grow into the parr, which have distinct dark bands on each side of their bodies. The non-migratory brown trout do not pass through a silvery smolt stage like the sea trout; instead they slowly develop the adult coloration.

Plate 44 shows a young trout with a fish leech attached to it. This leech lives by feeding on the blood of fishes, including trout and bullheads. *Piscicola* is instantly recognisable since unlike the other leeches in The Stream (page 88), it has a slender body with very conspicuous suckers. These suckers enable it to cling on to the

44. Small brown trout (*Salmo trutta fario*) with fish leech (*Piscicola geometra*) attached

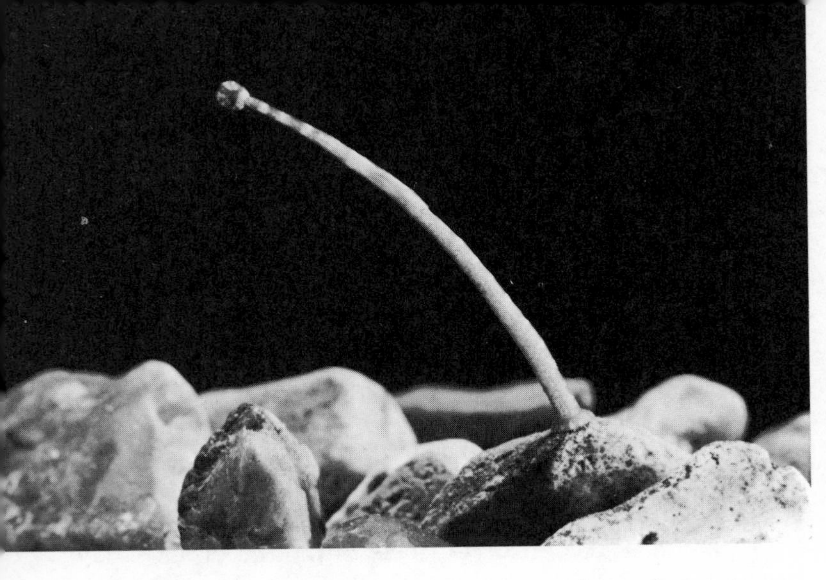

45.
Fish leech (*Piscicola geometra*) attached by hind sucker to stone waiting for fish

slippery flanks of fish and move up into the gills. The slender body also helps the leech not to be swept off by the fish's frantic effort to dislodge it.

At first when a fish leech is put into an aquarium, it loops its way over the bottom to a prominent site. There it rests attached by its rear sucker with its head end held out in the water (Plate 45). If it senses the movements of an approaching fish, it lashes to and fro, immediately attaching itself if it touches the passing fish. Since they are quite capable of killing small fish, fish leeches can be a menace in an aquarium.

6. Animals beside The Stream

(a) Recording the birds and mammals

Birds are identifiable both on sight and by their song. An outing with an experienced bird-watcher will soon teach you the field identification clues to the commoner birds along your stream. Songs can be learned by listening to one of the many bird song records that are now available (page 123). You can either make your own check-list of the birds you see on each visit or tick them off on one of the check-lists available from the RSPB and BTO (page 124).

A check-list regularly kept can be the basis for recording the arrival and departure of migrants. The date when singing starts often coincides with the start of nesting. The number of nesting pairs can be estimated in many species by the number of birds singing. Birds sing as a way of informing others of the same species that this is their nesting territory. Watch for other ways in which they defend their territory. They may have favourite singing perches. Where are they nesting? What materials do they use for nest-building?

Bird-feeding habits are also extremely interesting. This can be followed either by watching the birds themselves, or by collecting the pellets regurgitated by owls and herons. The pellets, if they are soaked in water, can be gently dissected and the remains of the prey identified, varying from feathers and elytra (wing cases) of beetles, to bones and skulls of small mammals.

Two species of birds may feed on the same sort of food, but in different places. To take insects as an example, some live on the willow trees, while others live on the low stream-side vegetation. The

46.
An enthusiastic young birdwatcher

food given to nestlings may be quite different from that eaten by the adults. A word of warning—do be *very* careful to keep a good distance away from the nest and observe with binoculars (Plate 46) so that there is no chance that the birds will be discouraged from feeding their young. If you find a rare bird nesting by your stream, leave it strictly alone and tell no one else about it, unless the nest is endangered.

Mammals are much harder to observe than birds. They have an acute sense of smell which birds lack, and so they have to be approached upwind. Many small mammals have poor eyesight and can be approached quite openly. But the slightest noise, like a snapping stick, will send them away. Perhaps the main reason why most people are totally ignorant of the mammals that live around them, is because many are nocturnal. But since most of them are insensitive to red light, they can be watched using a torch covered with red cellophane. When watching larger mammals, be as inconspicuous as possible. Wear subdued coloured clothing such as a camouflage jacket. By standing stock still, skilled observers can remain totally unnoticed by the animal so that it almost bumps into them.

An alternative method of studying mammals is to learn their signs and tracks. Soft mud at the stream edge will carry the prints of water voles and birds like moorhens (see Plate 54, p. 113). Patterns of feeding and types of burrow are often typical of individual species. They can then either be attracted by baiting them, or live-trapped in Longworth traps. Their sounds and calls can be recorded on a portable tape recorder. Many battery models are now available. Use a uni-directional microphone which picks up sound from one direction only, and if possible use a parabolic reflector. Once you start recording, you will be startled by the high level of extraneous noise man produces with aeroplanes, cars, motor-bikes—as well as pet dogs. This extraneous noise is the recordist's main problem. There is now a special society for wildlife sound recordists (page 124).

The basic techniques for photographing birds and mammals have already been outlined on pages 31–2.

(b) **Spring** March–April–May

On almost every visit to The Stream, I either saw or heard a moorhen (*Gallinula chloropus*) and a water vole (*Arvicola amphibius*) (Fig. 20).

Fig. 20 Water vole (*Arvicola amphibius*) feeding

The name moorhen is derived from 'Merehen', meaning a bird of wetlands. Moorhens are not found on moorland areas. The adult birds have a red patch on the forehead and a red beak with a yellow tip. The moorhen repeatedly flashes the white inverted V on the

47. Water vole (*Arvicola amphibius*) submerging

underside of its tail as it swims along in a jerky manner. The young birds are brownish with a yellow bill.

Moorhens are more at home swimming or walking. They can fly, but they need a long run along the water to take off. If alarmed, they can dive, but usually they swim frantically into the cover of reeds or emergent water plants. They can also sink into the water, until only the bill projects above the surface.

A moorhen's nest is built up on to a platform of dried water plants. Two or three broods of eggs are laid from March to August.

The water vole is also known as the water rat, but unlike the brown rat (which can also swim), it has a blunt nose and tiny ears. In fact, the ears merge in so well with the brown fur that they are difficult to see. The water vole has a long hairy tail. Very often the first sign of its presence will be the distinctive 'plop' as it dives from the bank into the water (Plate 47).

Water voles live in burrows along the banks of streams and rivers. The entrance is marked by a hole with a diameter of 6–8 centimetres. Around this hole all the plants—especially the reeds—are eaten down in a circular area by the water vole stretching out from its burrow to feed. This area is known as a water vole 'garden'.

The bulk of the water vole's diet is plants. Large plants are gnawed through their base and felled before being eaten. The voles lay in temporary food stores of piles of cut reeds which are stacked up on convenient rocks. More permanent stores are laid in underground for the Winter, since water voles do not hibernate. Their diet also includes water snails (*Limnaea* sp.), freshwater mussels (*Anodonta* sp.), fish and worms.

The voles are active throughout the day, but especially so at dusk. Each pair has its own territory. If a neighbouring male strays into another's territory, they fight, standing up on their hind legs and chattering their teeth. Early in Spring water voles build a large spherical nest from grasses and rushes. The nest is usually built inside the burrow, but on quiet stretches of water it may be built in vegetation near water above ground. About five young are born in April, and several litters can be produced by October.

A brilliant flash of blue was the first sight I had of a kingfisher flying upstream. On another visit, I was able to watch it more closely, sitting on a branch of the oak tree overhanging Kingfisher Deep (Fig. 21) ready to dive for fish. Although comparatively small (16 centimetres long), the kingfisher is one of the most spectacular British birds. It has a brilliant iridescent blue and emerald-green back, an orange chest, a white throat and bright red feet.

The large dagger-shaped bill is used both for feeding and for nest building. When the kingfisher dives it grabs a fish (usually a bullhead or a minnow) and flies up to a perch. The fish may be banged on the

Fig. 21 Kingfisher (*Alcedo atthis*) with bullhead.
It will turn the fish round before swallowing it head first.

perch to stun it, before it is swallowed head first. The usual reason given as to why the fish are swallowed head first is that it is to prevent the fish's scales sticking in the kingfisher's throat. But this cannot be true either for the bullhead which is scaleless, or for the minnow which has minute scales! However, if either fish was swallowed tail first, the operculum covering the gills behind the eyes and the paired pectoral fins would catch in the bird's throat. The kingfisher is quite capable of hovering before diving for his catch, if there is no suitable perch over the water.

Kingfishers nest in a tunnel in stream and river banks. In March, both birds begin flying at the bank, excavating a hole by loosening the soil with their bill. Once a ledge has been made, they can perch on it to dig in further. The tunnel rises slightly and ends in a circular nest chamber. Here the eggs are laid on bare earth. The kingfishers did not nest along my stretch of The Stream, although the farmer told me they had nested there in the past.

A prize-winning ciné film entitled *The Private Life of the Kingfisher*, made by Ron and Rose Eastman, revealed several interesting facts about the biology of the kingfisher. Sequences of kingfishers diving under water showed that they close their wings before entering the water and plunge towards the fish with their beak open. They make their way up to the surface by swimming with their wings. A sequence taken of the young birds in their burrow showed that instead of the young being given small pieces of fish, they ate whole fish almost as big as themselves. Later the chicks regurgitated the fish bones. A pile of fish bones is often a clue confirming that a hole has been used by kingfishers.

In April, swallows (*Hirundo rustica*) were seen gathering mud from The Stream banks for building their cup-shaped nests amongst rafters and on ledges of the nearby farm buildings. Each Spring, swallows return to Britain from their overwintering grounds in South Africa. They begin to arrive in March, and continue arriving until towards the end of April.

Swallows are dark blue above with a chestnut-coloured forehead and throat. A blue band separates the red throat from the rest of the underside, which is whitish. Swallows have a forked tail, with much longer tail streamers than their relatives, the house martins (*Delichon urbica*).

In Autumn, after producing two or occasionally three broods, swallows begin to assemble—often on telegraph wires—prior to making their southward migration to warmer winter quarters.

The first butterflies I saw flying in Spring were small tortoiseshells (*Aglais urticae*) and a brimstone (*Gonepteryx rhamni*). Both these butterflies hibernate over-winter as adults and reappear to feed on the early Spring flowers on sunny days. By April, they had been joined by orange tip butterflies. The orange tips on the forewings only occur in the male. The female has a black spot in the middle of each forewing. Orange tip butterflies lay their eggs on lady's smock and hedge garlic. I saw several in flight and spotted one laying on hedge garlic. The caterpillars resemble the seed pods on which they feed but sometimes they are cannabalistic—eating each other. They pupate and overwinter on the plants, and, like the seed pods, gradually turn brown.

On the 27th May, orange tip butterflies were still on the wing, as well as small whites (*Pieris rapae*) and wall browns (*Pararge megera*). As I climbed down the bank to photograph red campion (*Silene dioica*) growing beside the water along Hawthorn Bends, I noticed a group of small tortoiseshell caterpillars feeding on stinging nettles (Plate 48). The tortoiseshells lay their eggs in clusters, and even after

48.
Gregarious small tortoiseshell caterpillars (*Aglais urticae*) feeding on stinging nettle (*Urtica dioica*) in May

49.
A black and red frog-hopper (*Cercopis vulnerata*) on grass at Frog-hopper Patch in May

the caterpillars have hatched out, they remain clustered together. Lackey moth caterpillars (*Malacosoma neustria*) also show this gregarious behaviour, in contrast to the majority of other caterpillars which are solitary.

On the same day in May, I found masses of red and black frog-hoppers (Plate 49) on grasses, stinging nettles and codlins-and-cream on the bend of The Stream which I subsequently called Frog-hopper Patch. If the plants were knocked, they either dropped off or else made short flights to a neighbouring plant. Their flight comes to an abrupt halt as they suddenly furl their wings. A fortnight later very few of the frog-hoppers remained.

Their black and red bodies are an example of warning coloration which is used by many insects to advertise that they are distasteful to possible predators. The black and yellow caterpillars of the cinnabar moth (*Callimorpha jacobaeae*) and the red and black 7-spotted ladybirds (*Coccinella 7-punctata*) are other insects with warning coloration.

Unlike other frog-hoppers, *Cercopis* overwinters as a larva. It is the larval stages of frog-hoppers which produce the white froth—known as cuckoo spit—on many plants in early Summer. This froth is produced by the larva blowing bubbles into a liquid produced from its anus. The froth protects the larva from enemies and also from drying up.

Walking through the fields where the cattle had been grazing, I noticed that several of the cow pats had neat holes on the top. These holes indicate the presence of dung beetles. Sure enough, by prising the top of the pat off with a stick, I found several different kinds inside. There were black rove beetles (Staphylinids) with short wing cases, and oval-shaped beetles with a black head and thorax and brown or orange wing cases (*Aphodius* spp.).

(c) **Summer** June–July–August

The Friesian cattle grazed on the fields surrounding The Stream in rotation. One day in August I photographed several of them wading along The Stream to drink in the stretch leading up to Brooklime

50. Cattle drinking in The Stream near Brooklime Bend in Summer

Bend (Plate 50). Later in the year, after the farmer had voluntarily had his herd tested for brucellosis, all the drinking places were fenced off. This was to prevent the risk of infection to the cattle by the water-borne disease.

Throughout the Summer there was an abundance of insects. Orange soldier beetles (*Rhagonycha fulva*) were feeding on upright hedge parsley. These beetles are common in late Summer, pairing on various flat umbellifer heads, such as wild parsnip (*Pastinaca sativa*).

Wasps (*Vespula* sp.) were busy at the water figwort flowers. These small brownish-purple flowers are chiefly pollinated by the wasps. After alighting on the flower, they curve their abdomen in towards the stem in a characteristic manner (Plate 51). Also on water figwort occurred the curious rounded figwort weevils (*Cionus hortulanus*). These beetles have a pointed rostrum on the front of their head which can be tucked in beneath the thorax. Figwort weevils spend the whole of their life cycle on figwort or mullein. The larvae are brown and slime-covered, resembling little slugs feeding on the leaves. The slime is later hardened into a cocoon, inside which the larva pupates.

51.
Wasp (*Vespula* sp.)
feeding on water figwort
(*Scrophularia aquatica*) flower

52. (*left*) Small tortoiseshell butterfly (*Aglais urticae*) feeding on water mint flower (*Mentha aquatica*)
53. (*right*) Full-grown elephant hawk-moth caterpillar (*Deilephila elpenor*) feeding on codlins-and-cream (*Epilobium hirsutum*) in August

These rounded cocoons resemble the figwort fruit in shape and colour.

The flowers of both water mint and fleabane were very attractive to a great variety of insects in the late Summer. Small tortoiseshells, commas (*Polygonia c-album*) and a peacock (*Nymphalis io*) as well as meadow browns (*Maniola jurtina*), hover-flies and wasps were feeding on them (Plate 52).

One insect which even the most unobservant passerby could not fail to see was feeding on top of a codlins-and-cream spike late in the day in mid-August. The elephant hawk-moth caterpillar (*Deilephila elpenor*) is unmistakable (Plate 53). It has a horn projecting from its tail end which is typical of all hawk-moths. The overall colouring is brown, with a pair of conspicuous eye spots near the front. When feeding, the caterpillar extends its narrow head, which is somewhat reminiscent of an elephant's trunk—hence the common name. If the plant on which it is feeding is knocked, or the caterpillar touched, it immediately draws in its head, enlarging the segments with the eye-spots into a fearsome-looking 'head' which it jerks viciously from side to side. This sudden shock tactic is used to deter would-be predators.

Elephant hawk-moth caterpillars also feed on rosebay willowherb, and on fuchsias. Most gardeners immediately kill any of these 7-centimetre-long monsters they find devouring their prized fuchsias. It is a pity they do not transfer them to the nearest patch of willowherb, where they can pupate underground safely to emerge the next June as beautiful pink and bronzy-green moths.

Fig. 22 Cock reed bunting (*Emberiza schoeniclus*) in Summer

One day towards the end of Summer, a reed bunting (*Emberiza schoeniclus*) (Fig. 22) came to The Stream. These birds usually nest beside rivers and in marshy areas. When disturbed, both the hen and the cock birds divert attention away from their nest by shuffling along the ground, pretending that their wing is broken. Several other birds, such as the ostrich (*Struthio camelus*) adopt this method of protecting their chicks.

(d) **Autumn** September–October–November

Throughout most of the year—including Autumn—many golden-yellow dung flies (*Scatophaga* (=*Scopeuma*) *stercorarium*) jostled on the cow pats. The yellow furry flies are the males and they lie in wait for visiting females, which are smaller dusty grey-green non-furry flies. Even when alarmed, the males are reluctant to fly away and very soon return to the pat. But as soon as a female fly arrives, there is a frenzy of activity as the male flies immediately surround her and attempt to mate. The successful male may fly off with his mate, but the female returns later to lay her eggs. The larvae develop in the dung and pupate in the soil below.

These dung flies are carnivorous, catching other small flies which are attracted to dung or to hawthorn and bramble flowers.

Early in September a flock or charm of goldfinches (*Carduelis carduelis*) fed on clumps of spear thistles near Moorhen Meander and Rosebay Run. Honeybees (*Apis mellifera*) were alighting on the muddy areas to drink water in pools or at the water's edge. Near the oak tree, I disturbed a cock pheasant (*Phasianus colchicus*) resting beneath the shade of the willows.

Autumn is the time of bird migration. Early in the season swallows and house martins gather together before flying South. At the same time as the Summer visitors are leaving Britain, the Winter visitors begin to fly South from the far North. While some of these 'Winter' visitors may arrive as early as June, the majority arrive in Autumn.

(e) **Winter** December–January–February

Periods of cold weather drive flocks of redwings (*Turdus iliacus*) and fieldfares (*Turdus pilaris*) southwards. Both these birds were seen feeding in fields close to The Stream. Even many of the resident species gather into flocks foraging throughout the short hours of daylight. Without a constant supply of food, birds cannot maintain their body temperature and quickly die. Prolonged periods of snow and ice can cause catastrophic drops in the numbers of smaller birds in particular. After the long, hard Winter of 1947, the green woodpecker (*Picus viridis*) almost completely disappeared from our countryside. Similarly, during the extremely low temperatures of the 1962–3 Winter, when even water-mains were freezing deep in the ground, many birds, including kingfishers, died.

Mammals survive these conditions by hiding away in deep burrows well insulated from the cold above ground and eat their stores of food. Their main danger is being inundated by a flood and suffocated. Even so, a sprinkling of snow soon becomes covered with the pattern of tracks of mice and voles, proving that they do not hibernate during Winter. Winter is a good time for looking for bird tracks in both snow and mud (Plate 54).

The Winter days were not entirely drab and colourless. A grey

54. Moorhen (*Gallinula chloropus*) track in muddy margin in Winter

Fig. 23 Cock grey wagtail (*Motacilla cinerea*) in breeding dress

wagtail (*Motacilla cinerea*) (Fig. 23) took up residence near the road bridge. Its common name belittles its beautiful colouring. Even without binoculars its bright splash of yellow as it flies past, and the bob of its tail as it hops from stone to stone, are quite distinctive. Seeing it through binoculars for the first time, with its combination of bright yellow underparts with grey back and black tail, was quite breathtaking. In Summer the male's throat turns from its winter whiteness to black.

The moorhens were still in their favourite haunt, but then there were still plenty of water shrimps for them to feed on. Any insectivorous birds have to work hard in search of food on land. All I could find were a couple of ladybirds hibernating in the capsules of red campion. But mixed flocks of tits busied themselves searching the rough bark of the willows and the oak, so maybe a really keen-eyed searcher could have found more insects.

7. Conclusions

(a) Summary of field equipment

The following list of equipment is a summary of the items which have been mentioned elsewhere in this book. It is given as a guideline and it is by no means essential that you should obtain *all* these items. Certainly, it would be foolish to attempt to use them all at one time. Very often it will be possible to improvise; for example, a duffle coat or an anorak with large pockets can be used instead of a rucksack.

Remember that it is preferable to observe and to make notes rather than to collect lots of specimens.

General

Gumboots
Haversack
Field notebook and pencil
Camera with accessories
Hand lens
Polythene bags
Tins and specimen tubes

For studying plants

Hand lens
Scissors or secateurs
Polythene bags
Tins or match-boxes for seeds

For watching birds and mammals

Camouflage clothing
Binoculars
Check list of birds
Tape recorder

For sampling the water life

Water net
Grapnel
White or pale-coloured enamel or polythene dishes
Glass trough
Pair of tweezers or forceps
Pipette
Hand lens
Litmus paper
Screw-top jars, cream or yoghurt pots with lids
Waterproof black felt pen
Self-adhesive labels
Bucket

(b) Setting up a stream habitat

Many stream inhabitants are adapted to life in moving water. If they are kept in the still water of a small aquarium, they languish and may even die. Their lack of health may be due to a need for a constant high oxygen content in the water. This can be provided by using a small aerator powered from the mains. Such a pump is adequate for keeping many stream fishes including bullheads, minnows, sticklebacks and small perches (Plate 55). It is important to have a cover over the aquarium, not only to keep dust from getting in, but also to stop the fish from leaping out.

However, even in an aerated aquarium, some animals—especially mayfly nymphs and caddis fly larvae—do not survive well, because they need a flow of water. An obvious example is the net-spinning caddis larvae that weave silken nets across the current to trap their food. Stream conditions can be simulated indoors in various ways. Fig. 24 shows one simple set-up, in which a series of dishes is

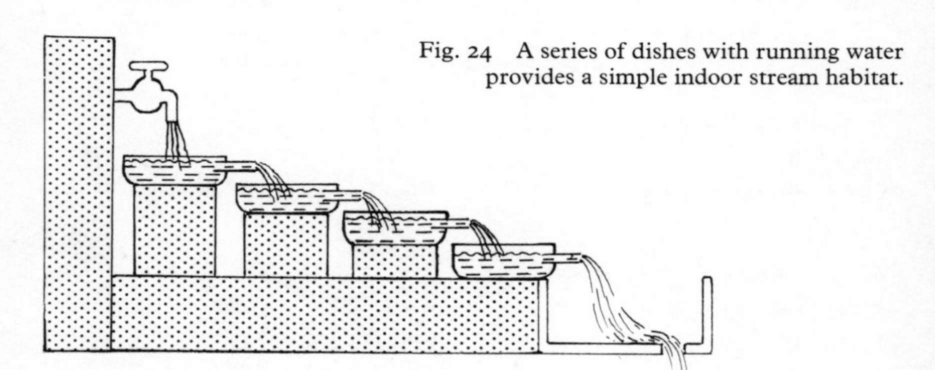

Fig. 24 A series of dishes with running water provides a simple indoor stream habitat.

55. A small perch (*Perca fluviatilis*) photographed head-on in an aquarium

arranged on a bench near a sink, so that each dish drains into the one below. If plastic dishes are used, the drain hole can be cut in the side and plastic or nylon tubing used for the outflow. The tubing can be sealed on to each dish by using an impact adhesive. Pieces of nylon tights or some other netting can be used to stop the animals from being washed through.

The flow of water can be produced by running water directly from a tap into the system. However, tap water is often heavily chlorinated to make it safe to drink. The chlorine in the water may either kill the stream animals directly or kill the microscopic animals and plants on which they feed. So water straight from the tap may prove to be lethal to your stream animals. If this is so, either collect some river or stream water, or allow some tap water to stand overnight and recirculate it from the lowest dish up to the top one by using a small electric pump. Peristaltic pumps are best, since the water does not come into contact with oily metal parts. A cheap and simple method is to use an aerator as an air lift. If a small funnel fitted with a piece of tubing is submerged inverted over the outlet of the aerator, the bubbles of air rise up the tube carrying small quantities of water with them, so long as the bore of the tubing is smaller than the bubbles the aerator produces. You may need to experiment with different-sized tubing to find the most efficient system.

Once the system is working a layer of river gravel and stones can be added to each dish, and an assortment of dead leaves and debris added to provide food for the detritus feeders. The stream habitat is now ready for its inhabitants. If you want to study the behaviour and life history of one species in detail, put only that type of larva in one of the dishes. Cover the dish with fine netting (nylon curtaining or tights) to prevent other animals from getting in. Since many stream animals are nocturnal they will be active only at night. Observe them by covering a torch with a layer of red cellophane and see if they react to this coloured light. Try to design systems that will give different current speeds, and see if different animals prefer slow or fast flowing water, and whether they use the eddies set up by obstructions such as stones in the water.

8. Further Reading and Information

Many more books have been written about pond life than stream life, but many of the inhabitants of ponds also live in streams. The books listed below will help those readers who want to read and learn more about the life which lives both in, and beside, streams. If any of these books are not in your own school library, then perhaps they can be borrowed or referred to at your local library or museum.

(a) Books for General Reading

These include books for readers who want to discover in more detail how to make a study of their own stream.

Angel, Heather, *Nature Photography: Its art and techniques*, Fountain Press/M.A.P., Kings Langley, 1972.

Angel, Heather, *Photographing Nature: Insects*, Fountain Press/Argus Books, Kings Langley, 1975.

Armstrong, P. H., *Discovering Ecology*, Shire Publications, Aylesbury, 1973.

Brown, V., *The Amateur Naturalist's Handbook*, Faber, London, 1960.

Clegg, J., *The Freshwater Life of the British Isles*, 4th edn., Warne, London, 1974.

Fitter, R. S. R., *Wildlife in Britain*, Penguin Books, Harmondsworth, 1963.

Fitter, R. and Fitter, M., *The Penguin Dictionary of British Natural History*, Penguin Books, Harmondsworth, 1967.

Flegg, J., *Discovering Bird Watching*, Shire Publications, Aylesbury, 1973.

Knight, Maxwell, *The Young Field Naturalists' Guide*, Bell, London, 1957.

Knight, Maxwell, *Field Work for Young Naturalists*, Bell, London, 1966.

Leadley Brown, Alison, *Ecology of Fresh Water*, Heinemann Educational, London, 1971.

Leutscher, A., *Field Natural History*, Bell, London, 1969.

Macan, T. T., and Worthington, E. B., *Life in Lakes and Rivers*, Collins, London, 1951.

Miles, P. M. and Miles, H. B., *Freshwater Ecology*, Hulton Educational, London, 1967.

Oldroyd, H., *Collecting, Preserving and Studying Insects*, 2nd. edn., Hutchinson, London, 1970.

Reade, W. and Stuttard, R. M. (eds.) *A Handbook for Naturalists*, Evans, London, 1968.

Sankey, J., *A Guide to Field Biology*, Longmans, London, 1958.

Seaward, M. R. D. (compiled by) *Advice for Young Naturalists*, 2nd. edn., Council for Nature, 1969.

Stephen, D., *A Guide to Watching Wild Life*, Collins, London, 1963.

Watson, G. C., *The Naturalist's Handbook*, Pan Books, London, 1973.

Whitten, D. G. A. with Brooks, J. R. V., *The Penguin Dictionary of Geology*, Penguin Books, Harmondsworth, 1972.

(b) Books for Identification

This list includes various reference books which are available for identifying the flowers, trees, freshwater life, fishes, insects, amphibians, reptiles, birds and mammals which can be found in and alongside streams.

Plants

Brightman, F. H. and Nicholson, B. E., *The Oxford Book of Flowerless Plants*, O.U.P., 1966.

Ellis, E. A., *Wild Flowers of the Waterways and Marshes*, Jarrold, Norwich, 1972.

Fitter R. S. R., Fitter, A. and Blamey, M., *The Wild Flowers of Britain and Northern Europe*, Collins, London, 1974.

Hubbard, C. E., *Grasses*, 2nd. edn., Penguin Books, Harmondsworth, 1968.

McClintock, D. and Fitter, R. S. R., *Pocket Guide to Wild Flowers*, Collins, London, 1956.

Martin, W. Keble, *The Concise British Flora in Colour*, Ebury Press, London, 1959.

Meikle, R. D., *British Trees and Shrubs*, Eyre and Spottiswoode, London, 1958.

Rose, F., *The Observer's Book of Grasses, Sedges and Rushes*, Warne, London, 1965.

Freshwater Life

The Freshwater Biological Association (see (d) p. 124) produces specialised keys for the identification of freshwater life. These include the following animal groups:

Flatworms (Triclads)—No. 23
Aquatic oligochaete worms—No. 22
Leeches (Hirudinea)—No. 14
Water fleas (Cladocera)—No. 5
Copepods (Cyclopid and Calanoid)—No. 18
Freshwater and brackish water crustaceans (Malacostraca)—No. 19.
Aquatic alder flies (Megaloptera) & lacewings (Neuroptera)—No. 8.
Black flies (Diptera, Simuliidae)—larvae, pupae and adults—No. 24.
Mayflies (Ephemeroptera)—nymphs—No. 20.
Mayflies (Ephemeroptera)—adults—No. 15.
Stoneflies (Plecoptera)—adults and nymphs—No. 17.
Adult Trichoptera (Caddis flies)—No. 28.
Water bugs (Hemiptera—Heteroptera)—No. 16.
Riffle beetles (Elminthidae)—larvae, pupae and adults—No. 26.
Fresh and brackish water snails (Gastropoda)—No. 13.
Freshwater fishes—No. 27.

The Royal Entomological Society (see (d) p. 124) also produces some specialised keys.

Clegg, J. (ed.), *Pond and Stream Life of Europe*, Blandford Press, London, 1963.

Corbet, P. S., Longfield, C. and Moore, N. W., *Dragonflies*, Collins, London, 1960.

Engelhardt, W., *The Young Specialist Looks at Pond Life*, Burke, London, 1964.

Janus, H., *The Young Specialist Looks at Land and Freshwater Molluscs*, Burke, London, 1965.

Macan, T. T., *A Guide to Freshwater Invertebrate Animals*, Longman, London, 1959.

Mellanby, H., *Animal Life in Fresh Water*, 6th edn., Methuen, London, 1963.

Needham, J. G. and Needham, P. R., *A Guide to the Study of Freshwater Biology*, Constable, London, 1964.

Southwood, T. R. E. and Leston, D., *Land and Water Bugs of the British Isles*, Warne, London, 1959.

Insects

Burton, John, *The Oxford Book of Insects*, O.U.P., 1968.

Chinery, M., *A Field Guide to the Insects of Britain and Northern Europe*, Collins, London, 1973.

Longfield, C., *Dragonflies of the British Isles*, Warne, London, 1949.

Riley, N. D. (ed.), *Insects in Colour*, Blandford, London, 1963.

Amphibians and Reptiles

Leutscher, A. (ed.), *Reptiles and Amphibians*, Blandford, London, 1962.

Leutscher, A. (ed.), *The Young Specialist Looks at Reptiles and Amphibians*, Burke, London, 1966.

Smith, M., *The British Amphibians and Reptiles*, Collins, London, 1951.

Fish

Maitland, P. S., *Key to British Freshwater Fishes*, Freshwater Biological Association Scientific Publication No. 27, 1972.

Muus, B. J. and Dahlstrom, P., *Freshwater Fish of Britain and Europe*, Collins, London, 1971.

Birds and Mammals

Bang, P., *Animal Tracks and Signs*, Collins, London, 1974.

Brink, H. van den, *Field Guide to Mammals of Britain and Europe*, Collins, London, 1965.

Fitter, R. S. R., *Collins Guide to Bird Watching*, Collins, London, 1963.

Heinzel, H., Fitter, R. and Parslow, J., *The Birds of Britain and Europe*, Collins, London, 1972.

Knight, Maxwell, *The Small Water Mammals*, Bodley Head, 1967.

Southern, H. N., *The Handbook of British Mammals*, Blackwell, Oxford, 1964.

Stehl, G. and Brohmer, P., ed. and trans. by Alfred Leutscher, *The Young Specialist Looks at Animals (Mammals)*, Burke, London, 1965.

(c) Bird Song Recordings

Quite often a passing bird will be heard but not seen. One way of identifying birds in the field is to learn to recognise their calls. The list of records given below includes birds which have been mentioned in this book. All these birds have been printed in heavy type. Mountain streams will have a quite different collection of birds.

Listen . . . the birds (Series available from the RSPB)

1. **Blackbird, song thrush, mistle thrush,** golden oriole, **robin**, cuckoo, nuthatch, **woodpigeon** and jay.
2. Garden and wood warblers, blackcap, **wren**, chiffchaff, firecrest, goldcrest, coal, marsh and great tits.
5. **Skylark**, woodlark, tree pipit, hoopoe, redstart, black redstart, **swallow** and sand martin.

6. **Starling**, pied flycatcher, serin, red-backed shrike, whinchat, **reed bunting**, nightjar and turtle dove.
7. **Dunnock**, house and tree sparrows, **greenfinch, goldfinch, chaffinch**, lesser redpoll, **linnet**, twite, corn bunting and **yellow hammer.**
11. Red grouse, blackcock, dunlin, wheatear, stonechat, ring ouzel, spotted flycatcher, pied, yellow and **grey wagtails** and corncrake.
12. Swift, house martin, **blue tit**, tree creeper, **carrion crow**, rook, **jackdaw**, **magpie**, kittiwake and Sandwich tern.
14. Sedge warbler, bearded tit, heron, bittern, red-necked phalarope, water rail, **moorhen**, coot, black-tailed and bar-tailed godwit.

(d) Useful Addresses

The British Trust for Ornithology (BTO)
Beech Grove, Tring, Hertfordshire HP23 5NR

Freshwater Biological Association (FBA)
The Ferry House, Far Sawrey, Ambleside, Cumbria LA22 0LP

Institute of Geological Sciences (IGS)
Geological Museum, Exhibition Road, South Kensington, London SW7 2DE

Royal Entomological Society
41 Queens Gate, London SW7

Royal Society for the Protection of Birds (RSPB)
The Lodge, Sandy, Bedfordshire SG19 2DL

Society for the Promotion of Nature Reserves (SPNR)
(Co-ordinators of the County Naturalists' Trusts)
The Manor House, Alford, Lincolnshire

Wildlife Sound Recording Society
Hon. Sec. D. T. Ireland, Esq., 1 Newcroft, Warton, Carnforth, Lancashire LA5 9QD

Index

English and scientific names of plants and animals mentioned in this book